Bones

Bones

The Life and Adventures of
Doctor Archibald Menzies

Graeme Menzies

WHITTLES PUBLISHING

Published by
Whittles Publishing Ltd,
Dunbeath,
Caithness, KW6 6EG,
Scotland, UK

www.whittlespublishing.com

© **2024** Graeme Menzies
ISBN 978-184995-591-1

Cover image, *The Grandeur of the Gardner Canal*, courtesy of John Horton, OBC, CSMA

Contents

Dedication

For Janet.
Though we know you least, you knew him best.

ACKNOWLEDGEMENTS

During my research, I found that many aspects of Archibald's life were reported but not necessarily proven; or in some cases they presented puzzles in need of resolution. In these instances I was very grateful for the patience and assistance provided by professionals such as Danielle Spittle, CRC Library Assistant at the University of Edinburgh Archives; Heather Kennedy from the University of Aberdeen; Leonie Paterson at the Royal Botanic Gardens Edinburgh Archives; Will Beharrell and Glen Benson of the Linnaean Society; Katherine Harrington, Mark Nesbitt, and Lynn Parker at Kew Gardens; Mrs JVS Wickenden, Historic Collections Librarian at the Institute of Naval Medicine; Ryan Cameron, India Rael Young, and Lisa Bengston at the Royal BC Museum Archives; and Emma Rutherford, Art Historian, Portrait Miniatures at Philip Mould & Company in London.

Scott McMaster at Castle Menzies, and Aberfeldy historian Tommy Pringle were also very helpful. I am also grateful to many members of the Menzies global diaspora for their support and encouragement including Sue Barnes in Australia, Ted Menzies and Baird Menzies in Canada, and Sean Menzies in the United States of America.

I am not sure if book-writing or publishing is an art, but I am grateful to several artists for their inspiration and talents. These include the late Jack Harman, OBC, whose sculpture of Archibald triggered my years-long research into Archibald's life, and his son Stephen Harman who so faithfully reproduced his father's work. The unparalleled marine paintings of John Horton, OBC, CSMA, similarly fuelled my imagination and interest in this area of research. I am indebted to John and Mary Horton for their unwavering support and enthusiasm, and grateful to have two of John's paintings reproduced in this book. The image on the cover, *The Grandeur of the Gardner Canal*, depicts open boats from HMS *Discovery* and *Chatham* surveying the west coast of what is now British Columbia – a view which has inspired me and which undoubtedly inspired Archibald as well.

Finally, all readers can join me in expressing gratitude to Martin Gavin for his editorial review of my original manuscript, and to Dr Keith Whittles of Whittles Publishing for crafting it all together into a work that, hopefully, you will enjoy.

Preface

In life it often occurs that a person becomes famous for one thing alone even though, like everyone else, their life and personality have multiple dimensions. With the passage of time, that one thing tends to dominate the memory of that person until all the other dimensions of their life become forgotten or remain completely unknown. This book reverses that phenomenon. It focuses on the aspects of a life which have been overlooked, or mentioned only in passing, and brings fresh perspectives and new dimensions to light.

This book is about a child of the Highlands and a product of the Scottish Enlightenment; he was a scientist and a philosopher; and he was also a naturalist and rugged outdoorsman. He achieved fame for being an accomplished botanist – indeed he was one of the very best – but for twenty years he was also a surgeon in the British Royal Navy, including service at one of the most celebrated naval battles prior to the Battle of Trafalgar. He circumnavigated the world twice, and for twenty years after his naval service was a celebrated doctor and raconteur in London. In these post-pandemic days, it is noteworthy that he was also one of the first experimental virologists. Finally, in the present day, it is also noteworthy that he was what we would call a progressive thinker: he recognised the sanctity of Indigenous faiths and condemned the evils of slavery.

The narrative of this story is chronological, although at times I have had to take some liberties. During his many months at sea Archibald was almost always on the move. He would be in Jamaica one month, Bermuda the next, then in Halifax, then New York, then Antigua. Later, with Captain Colnett then again with Captain Vancouver he would spend summers in what is now California, Washington, British Columbia, and Alaska, then pass winters in Hawaii, Maui, Kauai, before heading back to the Pacific coast of North America again. In these cases, rather than ask readers to ping-pong back and forth with the ships' itinerary, I have focussed the stories on where they happened rather than when they happened. I freely admit, for the sake of brevity, to having given short shrift to Archibald's experiences in California, Chile, and some other places.

Readers should also know I have taken some creative licence in the composition of this biography. It is not possible to know what every character in this story felt or thought in the various dramatic situations described so, for the sake of telling a compelling story, I have taken some liberties there.

The result of this work is, I hope, a more complete and wholesome record of a man's life and a better understanding of who he was as a person – not just what he accomplished for botanical and medical science.

Graeme Menzies
Vancouver, Canada

Maps and illustrations

North America, 1783. Map from the private collection of Roy Winkelman.

L'Amérique septentrionale, ou se remarquent les Etats Unis Paris, Chez Esnauts et Rapilly, 1783. Map.

Archibald's London. Map by author.

Croft Moraig. Photograph by author.

Castle Menzies, circa 1748. Watercolour by Paul Sandby.

Sir Robert Menzies, 1766. Painting by Sir John B. Medina.

John Hope, 1786. Etching by J. Kay.

Farmer George, 1786. Satirical print.

Joseph Banks. Painting by Benjamin West.

The Battle of the Saintes, 1782. Illustration by François Aimé Louis Dumoulin.

Royal Navy surgeon's tools. Photograph, Royal College of Physicians and Surgeons of Glasgow.

Botanist's vasculum. Photograph by Ji-Elle.

Chief Maquinna, 1791. Illustration by T. Suria.

Kwakwaka'wakw wolf mask. Photograph, UBC Museum of Anthropology.

Feather cape. Photograph, the British Museum.

Human bone sceptre. Photograph, The British Museum.

Canton factories, 1805. Painting by William Daniell.

British and Spanish ships at Nootka. Painting by John Horton.

Chief Maquinna dancing for Spanish and British delegations at Tahsis, 1792. Illustration by G. Gil.

Union hospital ship. Drawing, unattributed.

Miniature portrait of Archibald, 1803. Watercolour by Thomas Richmond.

Miniature portrait of Janet, 1803. Watercolour by Thomas Richmond.

Janet's fashion, 1798. Hand-coloured etching, Los Angeles County Museum of Art.

The Archibald Menzies Room at Castle Menzies. Photograph by author.

Archibald busts. Photograph by Stephen Harman.

One

STANDING STONES AND CASTLE GARDENS (1754–1770)

Aristotle is credited with saying, 'Give me a boy until he is seven and I will show you the man', and many still believe intuitively that life's earliest experiences shape us for the rest of our time. So, if we are to truly understand who Archibald Menzies was, we must begin our story in the Highlands of Scotland, near a group of neolithic standing stones known today as Croft Moraig. These mysterious stones, placed here by local inhabitants some five thousand years ago, were a landmark in Archibald's life. His family home where he was born was just, pardon the pun, a stone's throw away. His home has since disappeared, but Croft Moraig remains, visited today mostly by sheep. But a young Archibald, scrambling over them, might well have traced his fingers over their stone-tooled petroglyphs and wondered about the people who put them there, and what the stones and symbols meant to them.

Perhaps this was where his lifelong curiosity about people, and the world beyond Scotland, first took root.

Other things and places in the area around Croft Moraig would also have shaped young Archibald's mind and made an indelible mark on his character. The nearby town of Fortingall would have been one of them. Its church, initially a monastery, had been built around the year 700. He would have been fourteen years old when the belfry was installed in 1768 and may even have been present for the occasion. When visiting the church, he most certainly would have marvelled at its famous yew tree, which was also known to be at least a thousand years old.

A short walk from the ancient church and yew tree, Archibald would have seen a flat field with a bump on its profile. This was the Bronze Age burial mound called Càrn na Marbh (the mound of the dead), identified by a stone marker on top, Clach a' Phlaigh (the Plague Stone). According to legend, the stone marked the spot where the victims of a fourteenth century plague were delivered on a sledge drawn by a white horse led by an old woman.[1] It is still there today, in the middle of a potato field.

The burial mound, considered by some locals in the area to represent an entrance to the otherworld, is also where, once a year, people in Archibald's time and earlier had gathered to mark an ancient festival – Samhain – celebrating the end of summer and the beginning of winter. Here a bonfire giving off cleansing and protective powers was built, and at sunset, participants would dance around it in a circle – first one way and then the next. After the festivities were over many of Archibald's young friends would have taken a blazing torch from the fire and paraded it around the boundaries of their farms to protect the family from faeries and evil spirits. The new fire, kindled from the sacred communal blaze, was then traditionally brought into the house to light, heat, purify, and sanctify the home.

Wandering a little further east of the burial mound and plague stone, a rambunctious young Archibald might have seen another mysterious land feature: the remains of what some people said was once the site of an ancient praetorium built by a Roman general who lived there long ago. There was a local legend that Pontius Pilate, the governor of Judaea who presided over the trial of Jesus, was born here – the product of an exotic liaison between a Roman soldier and a local girl. Maybe this mound supported that theory.

Or maybe it was just a mound.

Other natural features in the area mingled with local tales, myths, and supernatural stories. The hillside forest at the village of Weem, just behind the presbyterian church where Archibald's baptism was certified in 1754, was home to an ancient mystical well named after Saint David. A cave in which another Menzies – Sir David – was said to have lived as a hermit around the year 1440 was also located there. Undoubtedly Archibald would also have hiked through the scenic woods of Moness, just a kilometre south of Weem, where he would have seen the birch trees and waterfalls that later inspired works by Scottish poet Robert Burns and the contemporary English musician Ed Sheeran.

The greatest natural feature in the area was Ben Lawyers, the tenth highest peak in all of Scotland. Climbing to its top through a path surrounded by ferns and heather would have provided the young Archibald with a physical challenge of stamina and would have rewarded him with panoramic view of the river, loch, and farming communities below.

Ancient tales connecting the people with the land, and the land with the stars, the seasons, and the Creator would all have been known to young Archibald, for many of the most ancient stories were part of an oral tradition passed down through the ages from one generation to the next. Tales of dragons, faeries, spirits, loch monsters, mystic symbols, sacred water, healing plants, and magical trees were the cultural backdrop to Archibald's youth.

In addition to these natural and supernatural influences, Archibald would have been profoundly shaped by the Highland clan system. He was, like so many in these lands, a member of a tribe.

Highland culture

The tribe that Archibald belonged to, the Menzies's, was one of fifty-two major Highland clans of Scotland and were not incomparable to many Indigenous tribes elsewhere in the world: the social hierarchy was formed by a shared view of heritage even though not everyone was biologically related. So, while many used the same surname not all were related by blood.

The Highland clans[2] that dominated Scotland, were an open-ended, inclusive, community firmly connected to the land. Inclusion in the clan was achieved by accepting and offering physical protection, land, political influence, and resources. The clan's leader, the chief, was the unit's leader and patriarch.[3] The chief was a warrior, a judge, a landowner, and the receiver and redistributor of resources. He was responsible for moral stature and reputation. He maintained the castle and church buildings, which were centres of commerce, law, faith, and social activity. In the region where Archibald grew up, many people signalled their identity with the clan by taking the chief's surname as their own. In other regions people used the term 'Mac' or 'Mc' to signal they shared a familial bond with the local chief. Likewise, each clan also developed its own distinct colours and patterns when weaving cloth – a tartan – to identify with the clan and to express membership. They had their own battle cries, their own origin stories, coats of arms, and legends.

The clan system that Archibald grew up in was not the political, social, and economic force that it had once been, but it was still influential. The castle gardens where Archibald, his father James, elder brother William, and other Menzies's laboured was part of a castle that dated back to 1500. Its predecessor, now a ruin located 5 km east, had been home to the chief since the 1200s. For 500 years, the land around Archibald's home – all that he could see in any direction – was ruled by the chief of the Menzies clan. The territory of the clan, like that of all other clans in the country, shifted constantly with the fortunes of the chief. At its peak, the Menzies chief controlled nearly 40,000 ha.[4]

Across this land for many generations Scottish Gaelic was the main language; and though educated and fluent in English, Archibald would have been proficient in Gaelic too. He would have been educated in, and familiar with, Latin as well. Though he could not have known at this time, his understanding of three languages and various dialects and accents would serve him well in later years. Because of this linguistic knowledge, Archibald would have known and understood the ancient Gaelic word *dùthchas*, a term difficult to define with precision but which some have described as an ecological principle that underscores and asserts a fundamental connection between people and the land.

Literally, the word *dùthchas* meant 'heritage' but it also meant much more than that. It also meant inheritance – including the right to live on the land occupied by one's forebears and it expressed the principle that clan members have a right to live in the land in the clan territory. It asserted the hereditary right of occupation and was at the heart of a clan's identity as a cohesive community with strong bonds of mutual obligation. *Dùthchas* was a rule of custom, not of law, and it was understood the clan chief was responsible for the distribution and regulation of the *dùthchas* in a benevolent and appropriate way. This meant chieftains were expected to recognise the needs and status of those living under their authority, and to protect and reward families that lived on and defended the *dùthchas*.

Another term Archibald would have known and understood was the concept known in Gaelic as *oighreachd*. This asserted the property rights of the chief over his land and over what his people could do on his land. This right of ownership – usually recognised officially by means of a feudal-style charter from the monarchs of Scotland – meant that clan chiefs were also landowners and landlords. They had major administrative and law and order responsibilities. The chiefs formed an important part of the Scottish landed gentry and were expected to uphold the power of the monarch and enforce the kingdom's law.

Since both of these terms were more custom and tradition than written law, developed over generations in a form of social contract rather than arguments in court by scholars and lawyers, the ideals of *dùthchas* and *oighreachd* existed in an ever-changing balance. But for the most part they complemented each other and gave cultural and social cohesion to clans, which enabled their chiefs to acquire, hold and expand their power as landlords.

Archibald would have understood that the combination of these principles made clans a highly effective form of social organisation and would carry these ideas and principles with him throughout his life and travels.

Like many Highlanders, Archibald would have felt profoundly that the landscape he occupied and observed functioned as a social, cultural, and spiritual map, rooting the communities that lived in them to their own history, culture, and distinct identity. His membership in the Menzies clan was more than blood, and more than fealty; it was firmly rooted in the land around him and all the people, plants, animals, stones, and spirits who were part of its organic, social, commercial, and political ecosystem.

Archibald's connection to the clan chief was not just traditional; it was also personal: his father James had been the head gardener at the castle for many years and was a trusted member of the castle estate.

Gardens of knowledge and nobility

James's role as head gardener was a substantial one; because the garden was not just a place where food could be grown but also a place of knowledge. James was responsible for overseeing the kitchen gardens, the physic (medicinal) garden, the pleasure garden, and the botanic garden. He was responsible for keeping all things – organic and otherwise – secure and in good order and repair; and he oversaw a workforce of twenty-one gardeners (half of whom also shared the same last name). The castle gardens were at least a century old – if not twice that – when James managed their development.[5] He also mentored youth including his sons, William, Robert, John, James, and Archibald, in the castle gardening profession.

Given what we know of his later life as a surgeon and doctor, Archibald was likely most interested in the castle's physic garden. The traditional healers in the clan would have told him about the ways plants could be used to heal people: that a dram of Columbine seed, given with a little saffron, could cure jaundice; wormwood would strengthen the stomach and liver; thyme juice mixed with vinegar would dissolve clotted blood; hydrangea root, mixed with other materials, could help women suffering from unusual dreams; and jasmine could be used to cure spots and pimples.

In addition to the influences of his father and brothers, Archibald would have been hugely affected by the actions of his clan's chief, Sir Robert Menzies. In his mid-sixties by the time Archibald was a teenager, the chief was a gentleman of the Enlightenment. His portrait in the castle shows him wearing a fine coat with brass buttons and a cotton shirt with lace cuffs; but it also shows he had large farmer's hands. He was undoubtedly a man of action as well as taste.

As chief, Sir Robert oversaw at an executive level the entire estate and was responsible for the welfare of the entire Menzies clan. He would have been pleased with the status of the lands around the castle – the area was well known for its excellent black cattle and thousands of sheep, and the fields produced good crops of oats, potatoes, peas, turnips, and flax. But the gardens at the castle would have been a place worthy of his special attention: the kitchen gardens supported the castle household and market for food, the physic garden provided medicines, the formal garden was aesthetically engaging, and the botanical garden was advancing knowledge and science.

Sir Robert attended to his gardens not merely out of duty, but also out of genuine personal interest. He was a keen naturalist, and his daily routine included recording the morning and evening temperatures and taking barometer readings. It was Sir Robert's decision, in 1747, to have a wall built at the north end of the castle's terraced garden so that fruit trees – plums, pears, and apples – could be planted against it. He also decided the garden should grow cherry trees – both black and red – which became widely regarded as among the best in the land. Years later, a Highland historian would write of the cherries:

Castle Menzies Gean [cherry] Trees from a very early date obtained celebrity all over the country. So famous indeed were these trees at one time, that their seed was sought from all parts of Great Britain, and large quantities were even sent abroad by request to all parts of the world. These Menzies gean trees were considered the oldest and finest gean trees extant; their fame has even been the poet's inspiration, there being an old Highland song composed in their praise, called 'The Castle Menzies Gean Trees', which was so popular in its time as to be sung all over Scotland. They also have been the musician's subject for composition – one of the finest dance tunes handed down to us is 'Castle Menzies Gein Trees Strethspey'.[6]

Following in the footsteps of James Menzies of Culdares, said to have brought the first larch tree seeds to Britain from Austria around 1717, Sir Robert also cultivated larch trees on the castle grounds.[7] He also grew elm, sycamore, and black Italian poplar trees. He was keen, like many other Scottish chiefs and English lords of the day, to obtain and grow plants from seeds obtained from around the world.

While Sir Robert may have been the ultimate gardener in Menzies territory, the greatest and most influential gardener in all of Britain was none other than King George III. Though in later years he was disparaged and mocked by revolutionary propagandists in America as 'mad King George', he was through most of his productive life known more fondly as 'Farmer George'. His interests were as sincere as they were serious: much of the nation's population was engaged in agriculture, and most of the nation's wealth was attached to agriculture too. The profitability of individual farmers was a legitimate concern of the state, since the wealth of the nation ultimately depended on their success.

Evidence of the king's interest in farming can be found in his library. It shows he studied the published works of noted agriculturalist Arthur Young and made extensive notes in his copies of *The Rural Oeconomy* (1770) and *The Farmer's Letters to the People of England* (1767). He also made notes, in French, on the French work *Le Socrates Rustique* ('The Rural Socrates') and copied detailed passages from a dictionary called *The Complete Farmer* (1766).[8] He even submitted original material to the *Annals of Agriculture* (edited by Arthur Young) under the pseudonym 'Ralph Robinson'.

Aside from reading and writing about gardens and agriculture, George III would go on to establish farms at Windsor, Kew, Richmond, and Mortlake, and conduct active hands-on research and practical experiments on his own estates. He followed with equally keen interest the innovations and practices of agriculturalist-technologist Jethro Tull[9] which, through use of science and technology, would rapidly increase British agricultural productivity. Between 1788 and 1811 he would gift 259 merino wool–producing sheep from his Spanish flock to 43 different recipients, thus improving the quality of British wool.[10] He was a patron of John Hope at the Royal Edinburgh Botanic Garden, supported Scottish botanist William Aiton and English adventurer-botanist-tycoon Joseph Banks in the transformation of the Gardens at Kew. He even built the world's largest telescope for astronomer William Herschel. He was madly progressive, and his enthusiastic support for scientific and agricultural discovery was boundless … and set an example that others were keen to follow.

Sir Robert was one of many in his social class who keenly followed George III's example and was eager to play his role in the advancement of science, agriculture, and knowledge. Sir Robert may have also taken a shine to the monarch since his brother-in-law, John Stuart the 3rd Earl of Bute, had been the young George's tutor.

In any event, Sir Robert combined his chiefly duty with Georgian sensibilities and sought a hands-on role in the affairs of his own estate and was personally engaged in the development of the castle gardens. Others of his social rank were doing the same. One hundred kilometres north of Castle Menzies, Sir Archibald Campbell of Cawdor Castle, which had been built centuries earlier around a legendary holly tree, created beautiful walled gardens for growing vegetables, fruit, and flowers. It still attracts tourists today.

The efforts of these land owners weren't limited to walled gardens. Landscape architecture also grabbed their attention and provided a way for the more enlightened nobles to express their connection with their lands, and they actively planned, pruned, ploughed, and perfected the acres of lands around their homes and castles.

In England, estate owners began to remodel their lands with picturesque vistas inspired by nature as the influence of Farmer George closed the curtain on the Renaissance era of opulence and ushered in the new Georgian era of the pastoral ideal. Formal pleasure grounds of wealthy estates, such as those at the recently completed Blenheim Palace, began to fade away as the new pastoral era began. This new landscape design movement used nature as its guide. Straight paths were replaced with meandering serpentine ones. Picturesque follies complemented artfully placed groves of trees to provide the country rambler with fresh vistas to delight the senses and new inspiration to fuel the soul.

Sir Robert undoubtedly learned of these developments, and of the new Georgian ethos, through his social contacts and business dealings in Edinburgh. A visit to the city in this period would have provided an opportunity to visit the Royal Abbey Garden at Holyrood House and to inspect the variety of vegetables and medicinal herbs grown there. Another destination would have been the physic garden at Trinity Hospital[11], which at one point had some 2,000 plants documented in *Hortus Medicus Edinburgensis*, the first botanical book published in Scotland (in 1683) by the first professor of botany at the University of Edinburgh, James Sutherland.

Edinburgh's new 2 ha Royal Botanic Garden would have also been a destination of choice. Its influential leader, John Hope, was the king's physician in Scotland, and a former student of the French naturalist Bernard de Jussieu at the Jardin royal des plantes médicinales in Paris.

At one of these Edinburgh trips, Sir Robert would exercise his chiefly duty to his people by recommending to Hope that he take teenage Archibald under his wing as a worker in the Royal Botanic Garden.

Scottish, and British

Before we review Archibald's time in Edinburgh, there's one further thing to know about his formative years: he was he was among the first generations of Scottish people to be raised British.

The 1707 union of Scotland and England was decades before Archibald's birth but it took another forty years for the dust from that momentous change to settle. Most people in Scotland had accepted the Union but there were others who felt that restoring the son of James Stuart (James II of England and simultaneously James VII of Scots) to the throne of Scotland would bring them more autonomy and freedom.

These proponents of the House of Stuart were the Jacobites – from *Jacobus,* the Latin for James. Undoubtedly young Archibald, like many other boys of his generation, would have been entertained and educated with stories of the Jacobite rebellion of 1715 which sought, but failed, to put James Francis Stuart on the throne. He may have been particularly amused by the story of John Campbell of Glenlyon who took Castle Menzies by force on behalf of the Jacobites in 1716 … only to be blockaded there until he surrendered.[12]

Although the 1715 rebellion had ended forty years before he was born, Archibald experienced the evidence and aftereffects daily. Two were unavoidable: the road, and the bridge.

British general George Wade had 420 km of road built after the rebellion to help keep an eye on the Jacobite-friendly districts in the Highlands and to increase the army's ability to advance quickly into the Highlands and tamp down any future uprisings. These roads, 16 ft wide and covered with loose gravel, fanned across the central Highlands connecting forts and barracks. One of them passed within 200 metres of Castle Menzies and the garden where James and his sons worked.

Wade also built several bridges. Archibald would have often have seen the one over the River Tay in his youth. Designed by architect William Adam, at a cost of some £4,000, the bridge was then, and perhaps still is, the most spectacular of Wade's 40 important bridges. Adams's humpbacked design featured a broad central arch, raised parapet, and four obelisks.

Like much architecture built after conquest, its purpose was both practical and political. It was a was both a bridge and a statement of power and authority. On its exterior, the central arch held relief carvings of a crown and crossed swords with the monogram of George II between them. The dedication panel mounted on the inner face, which Archibald would have read many times, held a Latin inscription that praised the ingenuity of the army and the benevolence and power of the king:

> Admire this military road stretching on this
> side and that 250 miles beyond the limits of the
> Roman one, mocking moors and bogs, opened
> up through rocks and over mountains, and, as
> you see, crossing the indignant Tay. This
> difficult work G. Wade, Commander-in-Chief of
> the Forces in Scotland, accomplished by his own
> skill and ten years labour of his soldiers in the
> year of the Christian Era, 1733. Behold how
> much avail the Royal auspices of George 2nd.

The principal strategic purpose of these roads and bridges was military: to provide troops based in the south with quick access to the eastern and central Highlands. Yet, ironically, and rather cleverly, they were used by rebellious Jacobites to march from north to south when they held their second uprising in 1745 – just a dozen years after the Wade bridge near Castle Menzies had been completed.

Growing up in the area after the rebellion of 1745, Archibald would have known that a few hundred Menzies warriors had participated in the Battle of Culloden in support of Charles Edward Stuart – also known as 'The Young Pretender' and 'Bonnie Prince Charlie'. He would have known that Prince Charles had marched down the Wade road and had stayed at Castle Menzies for two days, on his way to the battle against the government forces. Likewise, he would have known that some in the Menzies clan had joined ranks with the Atholl men on the Culloden battlefield, fighting in the notoriously vicious and visceral Highland charge and the fierce hand-to-hand fighting against the regiments of the government's front line led by the Duke of Cumberland.

Archibald would have heard the stories of how the musket balls and explosive mortar shells, fired by the government forces at close range, had devastated the proud, fierce, yet technologically inferior highlanders. In his future travels, Archibald would see this lesson of history repeated: highly skilled Indigenous warriors, no matter how brave and committed to the battle, were no match for an opponent with superior technology.

By the time Archibald was working at the castle gardens, the rebellion of 1745 was twenty years in the past. The men who had fought in that conflict were middle-aged. Archibald's generation was different. They knew the stories of the past, but they were not nostalgic for it. On the contrary. They were forward-looking and proud to be a part of the dawning new age of George III's Britain.

Archibald's generation could see that the union with England, though it had been difficult and at times bloody, was bringing prosperity to Scotland. Her commercial power, liberated and fuelled by trade with the American colonies and English markets to the south, was expanding daily. Wade's road system had doubled in size by the time Archibald turned fifteen and had more commercial than military use with each passing year. Castle Menzies was at peak performance, its gardens and other aspects booming, its chief never more prosperous. Thanks to the constant flow of traffic over the Wade bridge, the small town of Aberfeldy – just 2 km southeast of the castle and 6 km east of Archibald's home, not only survived these changes but continued to grow as a sustainable and proud focus for the agricultural and market community.

The 1793 edition of the *Statistical Account of the Parish of Dull* shows how much Archibald's home community had been transformed by the Union:

> The modes of dress and living in general have altered and improved within these last twenty years. No part of the old Highland dress is retained, except the Philabeg, 'the kilt', and the tartan hose. The coat or 'jacket' has short skirts. Great coats, or 'Highland cloaks' are now more generally used than plaids. The Sunday vests are commonly striped cotton; many of the young people wear watches. Many of the young women have printed cotton gowns and dusste cloaks, etc. 'Twenty years since' the people in this part of the country 'were universally jacobites; they are now, however, well affected to the present Government. The language spoken here is a corrupted dialect of the Gallic', and the natives of this part are acute and ingenious ... [13]

For intelligent and educated young men like Archibald, with the right patrons behind him, the opportunities for adventure and advancement were expanding exponentially and were in fact greater than he could imagine. Aware of the past, but grounded in the present, Archibald the young gardener left the ancient circle of standing stones at Croft Moraig and looked with great anticipation towards the future.

At age seventeen, under the patronage and blessings of his chief, he travelled to Edinburgh for the next chapter in his life, and the next in our tale.

Two

ENLIGHTENMENT, DISCOVERY, AND REVOLUTION (1770–1781)

The most formative years for many a young person, throughout the ages, are those between the age of seventeen and twenty-six. These years spent at university, and in first jobs, are when we meet the influencers and mentors who shape the trajectory of our professional lives. As fate and history would have it, these years of Archibald's life took place in a time and place that would be known for centuries as the centre of a great intellectual awakening: the Scottish Enlightenment.

Of this time even the great French writer-philosopher Voltaire was inspired to observe, 'It is a wonderful effect of the progress of the human mind, that today there comes to us from Scotland rules of taste in all the arts, from epic poetry to gardening'.[14] How fortunate for young Archibald to arrive in Edinburgh amidst this most extraordinary blossoming of knowledge.

Even more fortunate for him was that the position he took up at the Royal Botanical Gardens was supervised by one of the giants of the Scottish Enlightenment: Professor John Hope.

The professors

Forty-six-year-old John Hope was much more than a master gardener. He was also the King's Gardener and Keeper of the Royal Botanical Gardens Edinburgh; not only that, he was also a highly regarded professor of medicine and botany at the University of Edinburgh. The Edinburgh native had some pedigree as well: he was the son of distinguished surgeon and apothecary Robert Hope, known to the local community for volunteering his services at the Royal Infirmary.

After studies of his own, including time studying in Paris, John Hope was recognised, during Archibald's student days, as among the world's leading experts in botany. Indeed, his botanical lectures were the most advanced in Britain. He was a strong believer equally in the use of human senses – observation, smell, taste – and the use of reason to identify the unknown and to guide further discovery. He was a regular correspondent with Swedish botanist Carl Linnaeus, and a strong devotee of the taxonomy Linnaeus had developed for identifying plants, a method Hope had been using since 1766. The two exchanged letters, corresponding in Latin, sent each other seeds from newly identified plants, and often used their students as emissaries and couriers between Scotland and Sweden.

Hope was among the first to recognise the practical genius of Linnaeus's techniques, which established a precise yet flexible terminology of plant description, a stable and logical nomenclature, and a relatively simple method of plant classification. Dr Henry Noltie, a twentieth century Edinburgh botanist has said,

It is a measure of Hope's scientific understanding and its utilitarian application that he gave his students a thorough training in Linnaean methods long before their importance for botanical progress was generally appreciated. His teaching was responsible for the early adoption by almost all Scottish botanists of Linnaean nomenclature and descriptive terminology.[15]

This Linnaean nomenclature would be critically important as international trade increased. As new varieties and types of plants were found, the scientific community needed a common language to describe and identify them. Hope also collected many of Linnaeus's books (also written in Latin) at considerable expense and integrated Linnaeus's wisdom into his lectures.

All this knowledge was passed on to Archibald and Hope's other students.

But there was more to Hope than botany and Carl Linnaeus. As with many of his contemporaries, Hope had also been influenced by thinkers such as French philosopher René Descartes, English philosopher John Locke, and Scottish philosopher David Hume. Their influence on him was reflected in an aside to one of his students when he said,

It is the prerogative of man when he cannot be assisted by his senses to have recourse to his reason, and great are the discoveries that may be made this way, in this manner of investigation, our ground work must be a diligent observation of phenomena, and then let us exercise our reasoning faculty, taking care however to check a too exuberant fancy which might lead us astray.[16]

Skills of observation and of reasoning were the two great attributes Hope drilled into his students.

While many of Hope's students came from Scotland many others came from elsewhere. Thus, under Hope's instruction, Archibald was exposed to ideas and perspectives from a wide range of students from England, Wales, Ireland, and even further away: North America (predominantly Pennsylvania, Virginia, and the Carolinas), the West Indies (mostly St Kitts and Antigua), and continental Europe (including Switzerland, Russia, Germany, Denmark, Sweden, Holland, France, Italy, Spain, and Portugal).[17] Through these other students he learned of events, experiences, and places beyond his own knowledge and learned more about how the world was becoming increasingly interconnected in matters of trade and public affairs. Together they discussed innovations affecting the production and yield of tobacco plants in Virginia and sugar cane plants in the Caribbean; debated perspectives on the moral and ethical way those crops were produced; and discussed the competition between the major powers – Britain, France, and Spain – for dominion over those lands and crops.

Many of these conversations took place at the Royal Botanic Garden at Leith, just a kilometre or so north of the city – a unique teaching and research institution for Archibald and his international cohort of students. Hope's formal lectures took place on the upper floor of the gardener's cottage, and demonstrations of living plants took place outside in the garden, in its greenhouses, and at its Schola Botanica – an area dedicated to the study of plants of particular importance to medical students and those in the healing arts. For his services as a gardener – a role that was part labourer and part research assistant to the professor – Hope paid Archibald and other students a wage of 4 shillings and sixpence (about £ 20 in today's money) weekly.[18] It was a modest wage, but Hope also provided board at the Royal Botanical Gardens so, all things considered, Archibald was well situated. He had employment, lodgings, and the opportunity to work with one of the greatest and most influential surgeon-botanists of the day. The relationship blossomed, and in time Hope recognised his young student's talent and potential and actively took Archibald under his wing. He would not only be Archibald's employer and professor but eventually his patron, mentor, and friend.

Hope also encouraged Archibald to attend lectures at the university and introduced him to some of the world's most influential and brilliant professors.[19]

Archibald studied medicine and chemistry under Professor Joseph Black, one of the university's most popular lecturers and principal physician to King George III when he was in Scotland. Black's personal research was in the area of latent heat and, among other achievements, he discovered carbon dioxide and magnesium. Black was also a friend of fellow Scot James Watt, whom he had met many years earlier while both were in Glasgow, and had provided financial support to Watt's research on steam power.

In addition to Chemistry with Black, Archibald also studied anatomy under Professor Alexander Monro *Secundus*.[20] Monro was still a young man himself, thirty-eight years old when Archibald first attended his lectures in 1771, but was a leading expert: he had first attended anatomy lectures at the age of eleven, had studied in London, Paris, and Berlin, was equally skilled in Latin and Greek. He was the first to identify the lymphatic system as distinct from the circulatory system and, the same year Archibald arrived in Edinburgh, published a paper on the effect of drugs on the nervous system.

Archibald was also a student of Dr Francis Home, a former combat surgeon in the Seven Years' War, and a pioneer in the effort to vaccinate people against smallpox. Just five years before Archibald attended university in Edinburgh, Home was appointed the university's first professor of *materia medica* – a Latin term for 'medical material' – which was basically an 18th-century version of pharmacology. Home was also a clinical professor of medicine at the Royal Infirmary and, during the years Archibald was his student, president of the Royal College of Physicians of Edinburgh.

Another of Archibald's distinguished teachers was Dr Thomas Young, a professor of midwifery and the first to lecture on obstetrics; Young was also an accomplished surgeon and a pioneer in eye surgery, having devised a surgical knife for the removal of cataracts as early as 1756.

Archibald was also fortunate to be a student of the great Dr William Cullen – the sixty-one-year-old chemist and philosopher who would go down in history as a central figure in the Scottish Enlightenment. Cullen, a pioneer in artificial refrigeration, was a friend of Archibald's teacher and fellow chemist Joseph Black, physician and friend to philosopher David Hume and a professional and social colleague of many other leading figures of the Enlightenment such as Adam Smith and the enlightened aristocrat Lord Kames (noted for, among many other things, being one of the panel of judges in the Court of Session in Edinburgh who ruled in the 1771 Joseph Knight case that slavery could not exist in Scotland). Cullen was a professor of medicine, but his interest in, and appreciation for, botany were well-established: in 1754, the same year that Archibald was born, he had joined with Robert Hamilton, Regius Professor of Botany and Anatomy at the University of Glasgow, to advocate for a proper botanic garden, and a proper gardener to manage it 'in a manner becoming a Society devoted to Taste and Science, at the University of Glasgow'.[21]

Together, these and other professors at the University of Edinburgh provided young Archibald with access to the best academic instruction on medicine, chemistry, botany, surgery, pharmacology, maths, arts, and natural philosophy in the world.

He soaked it up.

Yet while Archibald was a keen and enthusiastic student, focussed on the studies that would lead him to a career as a surgeon and doctor, he also fully embraced life in the city: Edinburgh.

Edinburgh

Apart from what he learned at the Royal Botanic Garden, and in his studies at the university, Archibald was shaped by his immersion in the culture and surroundings of Edinburgh proper, with its magnificent castle towering above the city as a dominant and unmistakable landmark. The city, including surrounding areas of Canongate, St Cuthbert's, and Leith was a vibrant mix of people numbering some 70,000 souls engaged in all walks of life: nobles, intellectuals, priests, politicians, shoemakers, tailors, glove makers, smiths, saddlers, barbers, brewers, advocates, doctors, ship owners, artists, soldiers, printers, actors, poets, students, and others.

When walking from his lodgings at the Royal Botanic Garden towards central Edinburgh Archibald would have passed through an area known as the New Town which connected to the older part of the city – known subsequently as the Old Town. The New Town was the latest in urban planning, built to accommodate the city's growing middle-class merchant and professional population and constructed with the latest architectural styles and highest engineering standards. The New Town also reflected Edinburgh's enthusiasm for the progressive reign of George III through its street names: George Street, Queen Street, Hanover Street, Frederick Street (after George's father) and Princes Street (for his sons). Thistle Street and Rose Street expressed support for the still relatively new union between Scotland and England.

Though still under construction, Archibald would have enjoyed watching this new area of Edinburgh slowly take shape as he walked through it towards the Physic Garden and the new North Bridge linking the New Town with the Old Town.[22] Today the two are connected by a newer bridge, and the area is dominated by the Waverley Train Station.

Continuing over the bridge towards the old, smoky, overpopulated, and under-sanitised Old Town, Archibald may have passed near St Cecilia's Concert Hall and watched the city's musicians and music aficionados merging towards the hall's courtyard entrance on Niddry's Wynd. Still a relatively new structure when Archibald first saw it, the hall had been built by the Edinburgh Musical Society in 1763 and was the first purpose-built concert hall in Scotland, and the second in all of Britain. Archibald would never have seen anything like it before and likely thought its mere existence an exciting, resonating, symbol of the energy and sophistication of the city's inhabitants. The purpose-built oval concert room resonated with performances of composers like Wolfgang Amadeus Mozart, Joseph Haydn, and Johann Christian Bach.

Walking further south, past the Royal Infirmary on Robertson's Close, Archibald would have come to Nicholson Street, location of the Royal College of Surgeons and also the location of Colin Macfarquhar's print shop. Had he looked through the print shop window, Archibald may have gazed in wonder at samples of an ambitious new publication called the Encyclopaedia Britannica. The massive undertaking was a product of the city and the times – for there and then, knowledge was unfolding at a profound pace, and the dictionary was a physical manifestation of the Scottish

Enlightenment. Even the printer himself, son of a wig-maker, was an active participant in the movement, dedicating the first two sentences of his preface to the first edition to the idea that 'Utility ought to be the principal intention of every publication' and that 'Wherever this intention does not plainly appear, neither the book nor their authors have the smallest claim to the approbation of mankind'.

The third of the Encyclopaedia's initial three volumes was published the year Archibald arrived in Edinburgh and its thousands of entries and many engravings fuelled and inspired his cravings for an active role in further discovery.

Had Archibald rambled further east down Nicholson Street he would have seen on the horizon one of Edinburgh's most unmistakable features: a large volcanic hill known as Arthur's Seat, the focal point of Holyrood Park. Climbing to its 251 metre pinnacle was nothing compared to the hills and terrain around Castle Menzies, but would have rewarded Archibald with a fine view of Trinity College Church's gothic architecture, and the physic garden he would have passed on his way into town.

Arthur's Seat was something of an anachronism to the intellectual enlightenment unfolding below. An old legend said the hill was actually a dragon which, having feasted too enthusiastically on local livestock had lain down for a post-feast snooze never to wake up again. Others held that the hill was the site of Camelot, regal residence of the legendary King Arthur. Nevertheless, it was a prime spot to look down on the city; and it inspired not only fanciful legends but also works of art, including some noteworthy poetry. Archibald may have been amused by the mention of Arthur's Seat in Robert Fergusson's lengthy 1773 poem 'Auld Reekie' (Old Smoky – a nickname for Edinburgh, based on its often-smoggy appearance caused by the coal and peat fires from the cramped Old Town).

The philosophers

In his urban rambles, and as part of his student life, Archibald would have passed by and participated in many unofficial centres of learning that would shape his character, mould his mind, and form a reference point through which he would interpret much of his future life.

These unofficial centres were the social clubs and societies formed by the city's intellectuals, including Archibald's professor, and mentor, John Hope. Hope was a founding member, in 1773, of the Aesculapian Club – a dining club designed to promote goodwill and fellowship between physicians and surgeons; in later years when Archibald was a more established student Hope likely invited him to join their monthly meetings at various taverns around the Old Town.[23] The Club had twenty-two official members – eleven physicians and eleven surgeons – and met 'for the purpose of wining and dining together under the peculiar care and patronage of Apollo as the god of poetry and music, and Bacchus as the god of wine, and Venus because the club at first met on Friday – *die Veneris*'.[24]

The Aesculapian Club, and others like it, were an excellent, lively, and entertaining way for Archibald to meet informally with his other professors, other leading thinkers and personalities, and other bright students of the day. Hope also had a founding role in the establishment of the Newtonian Club in 1778, inviting Encyclopaedia printer William Smellie to serve as the club secretary, and by his example Hope encouraged promising students like Archibald to extend their professional and academic interests into their social lives.

Archibald's chemistry professor, Joseph Black, was a member of the more serious and more prestigious Poker Club, intended to poke ideas around in much the same way a fireplace poker pokes coal. Here, Black shared company and ideas with many of the city's intellectual heavyweights like philosopher David Hume; economist-philosopher Adam Smith; Church leader Alexander 'Jupiter' Carlyle; professor of natural philosophy John Robinson; Presbyterian minister, royal chaplain to George III, and principal of the university, William Robertson; Law professor Henry Dundas, who was made Lord Advocate (the top legal authority for Scotland) in 1775 and as counsel to former slave Joseph Knight concluded his remarks in that landmark case by stating, 'Human nature, my Lords, spurns at the thought of slavery among any part of our species', unequivocally establishing that any slaves domiciled in Scotland were in fact free, and also effectively abolishing domestic serfdom.

Another Poker Club member, merchant John Clerk of Eldin, had radical ideas regarding the way naval war was conducted and shared his thoughts with club members, eventually putting his ideas to paper in a work entitled *An Inquiry into Naval Tactics*.

These ideas about human nature, economics, justice, and even the conduct of war, were revolutionary (though not in the treasonous sense) and shaped the minds of the professors and in turn, their students.

Though he benefitted from the trickle-down effect of the Poker Club, Archibald was not yet of sufficient accomplishment or stature to be a member; however, The Philosophical Society was pleased to have him as a member.[25] The society had been established in 1737 by Alexander Monro *Primus*, father of Archibald's anatomy teacher, and five other men, including giants of the Scottish Enlightenment like David Hume and James Douglas, 14th Earl of Morton.[26] Like other Enlightenment figures, Morton elevated reason and science as the most rational method for achieving knowledge. In keeping with this view, Morton was committed to the philosophy of securing greater autonomy and freedom for ordinary people.

One of Morton's suggestions (often referred to as a 'hints') to Captain James Cook prior to his first voyage in 1768, reflected a philosophical approach that Archibald would take to heart on his own voyages around the world. Morton wrote that Indigenous people were created equal, and had rights:

> They are human creatures, the work of the same omnipotent Author, equally under his care with the most polished European; perhaps being less offensive, more entitled to his favor. They are the natural, and in the strictest sense of the word, the legal possessors of the several Regions they inhabit. No European Nation has a right to occupy any part of their country, or settle among them without their voluntary consent. Conquest over such people can give no just title; because they could never be the Aggressors.[27]

Morton's position aligned with George III's 1763 royal proclamation protecting Indigenous property rights in North America. That proclamation, issued in October of that year, addressed the conflicts and tension on the eastern side of the North American continent by closing down further colonial expansion westward and attempted to shield the Indigenous people from attacks by settlers. Colonial governments and private citizens were forbidden from buying land or making any agreements with the Indigenous peoples (the Crown would handle that, if and when necessary) and only licensed traders would be allowed to travel west or deal with Indians. George III's proclamation read in part:

> And We do hereby strictly forbid, on Pain of our Displeasure, all our loving Subjects from making any Purchases or Settlements whatever, or taking Possession of any of the Lands above reserved without our especial leave and Licence for that Purpose first obtained.[28]

This proclamation, the Morton hints, and other topics – ranging from developments and theories in medicine, surgery, botany, and chemistry to those of philosophy and politics and beyond – fuelled lively discussions at the Edinburgh clubs and societies. What's more, since people belonged to several clubs at the same time, the ideas discussed one night with one group were often repeated and discussed again a following night at a different club, with each iteration being further defined and polished. The ideas spread equally through publication and through word of mouth.

The intellectually lively Edinburgh scene was such that it inevitably attracted American Benjamin Franklin (the eminent inventor-politician) to visit, arriving in October 1771 not long after young Archibald had settled in. Franklin stayed with David Hume at his home in New Town and over his two-week stay renewed his old acquaintance with Lord Kames, with whom he had become friendly during his previous visit to Scotland, and with many of Archibald's professors and associates, concluding with the observation that '… the University of Edinburgh possesses a set of truly great men, professors of several branches of knowledge, as have ever appeared in any country'.[29]

What good fortune, and how incredibly stimulating, for Archibald to be a student in Edinburgh then. He received an education that was, at that time, truly second to none.

Discovery

While the many great advances in medicine, natural science, and philosophy taking place in Edinburgh captivated Archibald's attention, external events also fed his imagination.

In another great gift of fate, the year he arrived in Edinburgh was the same year forty-three-year-old Captain James Cook completed his first voyage around the world. News of this great mission rapidly found its way to Edinburgh and especially to Hope and his eager students. The voyage had been a scientific one that included several important civilians: astronomer Charles Green (funded by Lord Kames) and Hope's close associate the botanist Sir Joseph Banks (self-funded), who was also accompanied by the distinguished Swedish naturalist (and the ship's surgeon) Daniel Solander.

Hope and his students, along with many natural scientists in Britain elsewhere, were electrified by Banks's account of his voyage with Cook. He returned from the voyage with no fewer than 30,000 plant specimens, catalogued using the Linnaean taxonomy favoured by Hope and his protégés. In next to no time, Banks became Britain's most popular and celebrated botanist, a much-sought-after speaker, and his status as botanical adviser to King George III was permanently cemented.

Banks had been just twenty-five years old when he set out with Cook on HMS *Endeavour*. The intelligent, wealthy, adventurer-aristocrat was one of 94 men and boys on the ship. Cushioned by an annual income of £5,000 (about £500,000 a year, today), Banks was unconcerned about the relatively small size of the ship – just 30 metres long – and brought with him a personal secretary, two artists, four servants, and two greyhound dogs.

Aside from his onshore expeditions in Rio de Janeiro, where he documented the first scientific description of the Bougainvillea plant, and in Tahiti and the east coast of Australia, Banks was most enraptured by the six months he spent with Cook in New Zealand. In addition to his many botanical explorations and discoveries, and the astrological work on the transit of Venus, he was fascinated by the Indigenous peoples' use of a hand-held weapon known today as a *patu onewa*. He collected some examples through trade, and in his journal made sketches of others and described them as follows:

> … a kind of small hand bludgeon of stone, bone or hard wood most admirably calculated for the cracking of sculls; they are of different shapes, some like an old fashioned chopping knife, others of this [sketch] or [sketch] always however having sharp edges and a sufficient weight to make a second blow unnecessary if the first takes place; in these they seemed to put their cheif dépendance, fastning them by a strong strap to their wrists least they should be wrenched from them. The principal people seldom stirred out without one of them sticking in his girdle, generaly made of Bone (of Whales as they told us) or of coarse black Jasper very hard, insomuch that we were almost led to conclude that in peace as well as war they wore them as a warlike ornament in the same manner as we Europeans wear swords.[30]

Anticipating a return to the area on Cook's second voyage, due to set sail the next year, Banks quickly commissioned several brass reproductions of the *patu onewas* from the samples he had collected, so that they could be presented as gifts – in the manner Europeans would present elaborate ceremonial swords as gifts – to local chiefs and dignitaries and others of high social status.

The impact of Cook's voyages – this first and the two others that would conclude in 1775 and 1780 – are difficult to underestimate; in the broader world they were of such epic proportions they would be talked about and analysed for centuries. But the immediate effect on Archibald was profound and life-changing. He was impressed and encouraged to know that Cook was the son of a farmhand who, after joining the Royal Navy at the relatively old age of twenty-six, rose to fame through his own intelligence, determination, and merit. In this progressive Georgian era, one didn't have to be rich and titled to achieve greatness.

The reports of Cook's discoveries and adventures, which were a constant backdrop to Archibald's life as a student in Edinburgh, also encouraged him to consider a career in the Royal Navy. As a botanist like Banks, or as a surgeon-botanist like Solander, he could travel around the world, visit unimaginable places, meet new people, and participate in the discovery of things unknown to science.

It was not a pipe dream. Cook's third voyage included a month-long visit to a place on the uncharted north-west coast of North America called Nootka Sound.[31] The Nootka area was populated by the Nuu-cha-nulth peoples, and had been their homeland for some 4,000 years.[32] Although Spanish Captain Juan Pérez had visited Nootka Sound previously, in August of 1774, on his ship *Santiago*, Cook was the first European to set foot at Nootka when he landed there in 1778. This seemed a small matter at the time, but later it would be an important detail – one that would almost result in war, and which certainly set in play a series of events that would give Archibald the opportunity of his dreams.

While the details of his future and the impact these current events would have upon them were unclear to Archibald at this time, he surely understood that he would have to work hard if he were ever to follow in Banks's or Solander's wake. The Cook expeditions undoubtedly motivated him to be both an accomplished botanist and an excellent Royal Navy surgeon.

Revolution

Thrilling as the tales of Cook's voyages were, they competed for Archibald's attention with the stories and discussions regarding the status of the colonies in North America.

Before arriving in Edinburgh Archibald understood that England's union with Scotland in 1707 had taken some decades to settle in, and the Union had since experienced two unsuccessful Jacobite rebellions, with the financial and military aid of France. He could see the same pattern and tensions unfolding in North America as both Britain and France competed for control of their own – and each other's – new territories there.

George III's Proclamation of 1763, issued at the conclusion of the Treaty of Paris, was the first of several efforts to impose greater control over its colonies, and its prohibition on further colonial expansion westward was an irritant to the Americans. The Sugar Act of 1764 and the Stamp Act of 1765 – new taxes levied to help the government recover the costs of defending its colonies from French control – was a further irritant.

In March of 1770, the year before Archibald's arrival in Edinburgh, violence in Boston – characterised as 'The Boston Massacre' – saw five civilians shot by government forces, which energised anti-government sentiment among the colonists. During his second year at university, Archibald read reports of the 'Boston Tea Party'. By the time he was into his fourth and fifth year in Edinburgh armed conflict between the ambitious unfettered settlers and the restrictive and heavy hand of government had escalated in both frequency and in scale. Copies of Thomas Paine's 47-page work *Common Sense*, published in Philadelphia in 1776, fuelled the cause of self-rule. They also made their way to the social clubs of Edinburgh, further stimulating discussions about the state of nature, the role of society and government, and the role of monarchy. Likewise, the publication of the Declaration of Independence document, issued later that same year by members of the Second Continental Congress, which included Hume and Kames's old friend Benjamin Franklin, also generated much discussion around the philosophical and other societies in Edinburgh.

For the time being, Archibald was not directly affected by the tensions on the east coast of North America. But he was aware of the issues and the events, and in time, they would affect him personally.

Field work

The vibrant city, the nights philosophising, the adventures and discoveries of Cook, and the emerging political issues in America, were the backbeat to Archibald's academic work, which continued apace. But not all his academic work was performed in Edinburgh. He was also field-tested.

As Archibald was one of his top students, John Hope arranged for him to undertake several special field assignments. One was to collect plants from the alpine heights of Britain's highest mountain, Ben Nevis.[33] Another was a botanical field trip to the Hebrides. Both of these challenging missions were important commissions for Hope's distinguished and influential gardening associates in England: Dr William Pitcairn and Dr John Fothergill.[34] Pitcairn was president of the Royal College of Physicians of London, and Fothergill a former student at Edinburgh University, a friend of Benjamin Franklin, and a highly respected London physician.

The two physicians were also avid gardener-botanists who had been inspired by recent voyages to the Hebrides by such as Joseph Banks and the Welsh antiquarian, geologist, and naturalist, Thomas Pennant.

Banks had toured the Hebrides after his much-celebrated return from the Pacific with Cook, but not before meeting with Pennant to present him with the skin of a new species of penguin he had collected from the Falkland Islands. Energised by Banks's interest in the Hebrides, Pennant visited the rugged west coast islands shortly afterwards, stopping at Bute, Arran, Ailsa Craig, the Mull of Kintyre, and Gigha. Like Hope and Banks, Pennant was a devotee and lifelong correspondent of Carl Linnaeus, and was indebted to Linnaeus for nominating him for membership in the Royal Swedish Society of Sciences following his early scientific papers on geology and palaeontology.

Pitcairn and Fothergill were both too old and too busy to undertake a voyage to the Hebrides themselves, but they had extensive botanical gardens and were keen to improve their knowledge of plants within the British Isles. Pitcairn's garden at his Islington home was 2 ha in size; Fothergill's in Upton, a massive 24 ha. Of Fothergill's garden Banks would say, 'In my opinion no other garden in Europe, royal or of a subject, had so many scarce and valuable plants. It was second only to Kew in attracting visitors from overseas.'[35]

Archibald's success in the Hebrides, capably following in the footsteps of established names like Banks and Pennant, satisfied his mentor and made a good impression on two well-regarded and influential surgeons and botanists.

On his Hebrides mission he not only collected plant specimens for his patrons, he also collected valuable experience travelling over land and sea, confirming to himself and to his patrons that he was indeed a capable adventurer. The hours at sea, landing here and there for explorations by small boat, navigating the rocky coastline, and gathering stories of ancient folklore, were a vastly different experience from his urban lectures; but these experiences likely gave him confidence that everything he had learned in the previous few years had prepared him for a future of adventure and discovery.

After nearly ten years in Edinburgh, his mind and natural abilities had been expertly cultivated by his professors, and he was now ready for the next stage in his life.

On 8 October 1778 Archibald received his qualification as a surgeon in the Royal Navy.[36] It was an ominous time to take this important, life-changing step: four months earlier, France had escalated its support for the revolution in America by sending a fleet of 12 battle ships to blockade the British navy fleet in the Delaware River; and, four months hence, the great and capable Captain Cook would be cut down on a beach in Hawaii.

These distant events would impact Archibald's life in ways he could not yet imagine.

Three

Blood and botany (1781–1786)

Archibald would have been proud to qualify as a surgeon in the Royal Navy and eager to begin his practice at sea. He would have been equally excited to see the world beyond the British Isles and keen to leverage his botanical skills to that end. But, like many a young graduate, he likely found the first two years following his qualification to be tedious and uneventful.

Leaving Scotland, he took a position in Caernarfon, Wales, as assistant surgeon and waited for his opportunity to join a ship.[37] Eventually his patience paid off and, qualified both in surgery and physic, he was ordered to Portsmouth where he would take his first assignment as ship's surgeon.

Portsmouth was the perfect place to wait for his assignment, for it was both educational and inspirational. The city's relationship to the navy and shipbuilding dated back at least 500 years – its natural harbour and proximity to the Normandy coastline just 125 km to the south made it the perfect place to gather troops and ships in preparation for an invasion of France. By the 1400s it had become more permanently fortified to protect it from counter-attack, and by 1700 it was one of the most fortified cities in the world and also a centre of ship manufacture and innovation. From here, General James Wolfe sailed to capture Quebec in 1759, and Captain Cook first set off to circumnavigate the globe aboard HMS *Endeavour* in 1768.

Many great adventures began here; and news of events, places, and people from far away arrived here first. It was an exciting place to be, to leave, and to return to.

In recent years a new hospital had been built in the area: the Royal Haslar Hospital. As a medical man, Archibald used some of his time to investigate the massive three-sided building which had reportedly cost £ 100,000 to build and another £ 14,000 annually to maintain. The governor of the hospital was physician James Lind, a Fellow of the Royal College of Physicians of Edinburgh, supported by two surgeons, an apothecary responsible for the distribution of drugs, seven surgeon's mates, three further assistants, and a large number of female nurses.[38] It was the largest and most modern hospital Archibald had ever seen, and he took full advantage of his time in Portsmouth to visit the hospital and learn of the latest advances and theories with regard to the injuries and ailments suffered by the many hundreds of seamen treated there.

Eventually, Archibald's ship came in. Literally.

HMS *Nonsuch* was a beauty: she was 49 metres long and had twenty-six 24-pounder guns[39] on the lower gun deck, twenty-six 18-pounders on the upper deck, ten 9-pounders on the quarterdeck (rear), two 9-pounders in the forecastle (front), and a crew of just under 500 souls. Although relatively new – she had been built here at the Plymouth dockyard just eight years earlier – she and her crew had seen considerable action. She had been to America and, the year before Archibald joined her, had participated in the successful effort by the Channel Fleet to relieve Gibraltar from its opportunistic siege by Spain (with support from France).

On the way back to the Channel from Gibraltar the *Nonsuch's* plucky crew, still rejoicing from their victory, engaged the mightier seventy-four-gun French ship *Actif* as she harassed and preyed upon British ships from her haven in Brest on the west coast of France, 200 km south of Portsmouth. *Nonsuch* then returned to Portsmouth for repairs and a change in crew – which now included Archibald.

The ship had lost 26 men and had 64 wounded in the action against *Actif*, so Archibald fully expected his skills as a surgeon would be put to good use in the weeks and months ahead. The ship also received a new commander, Captain William Truscott. This was the eighth ship for the seasoned forty-seven-year-old Truscott, who took command of *Nonsuch* on 15 August 1781 shortly after participating in a North Sea engagement against the Dutch, who had been providing support to the revolutionaries in the American colonies earlier that month.[40] They left Portsmouth to patrol the Channel, ever ready to engage the Dutch, the French, or the Spanish.

As surgeon, Archibald's status on-ship was equivalent to that of a warrant officer, like the master and the purser, and as such was one rung down the status ladder below the commissioned officer class.[41] This meant he did not wear an officer's blue coat, and did not dine in the wardroom (the officers' mess) by right … though he would have done so by standing invitation. Uniforms in the British navy had only been formally introduced 30 years earlier and were only expected to be worn by commissioned officers. Surgeons like Archibald, and others of warrant-officer status, dressed in upper-civilian-class dress of their choice: trousers or knee breeches and stockings, black shoes, a waistcoat, linen shirt, wool full coat, and a cocked hat (sometimes called a tricorn hat).

Fortunately, surgeons were assigned a personal cabin off the gunroom and spared the indignity of having to bunk with the younger and much rougher crew; so in this regard at least, he was one rung higher up the status ladder than the midshipmen and two above the regular seamen.

To prepare for his role, Archibald would have studied the somewhat antiquated *Regulations and Instructions Relating to His Majesty's Service at Sea*, first published in 1731. The manual explained the surgeon's responsibility to visit the sick twice a day, present a sick list to the captain daily, and maintain a log of activities. He had full authority and responsibility for treating minor ailments and injuries on the spot and could also move patients to the sick berth – which, depending on the ship and the circumstances, might simply be a canvas cubicle rigged up between two gun ports. Any given day might bring him seamen with broken limbs, muscle strains, ankle sprains, skin sores, and burns. He had to watch for epidemics of dysentery, typhus, scurvy, malaria, and yellow fever, and would also be expected to treat crew that had suffered corporal punishment (flogging).

In battle, he was expected to relocate to the orlop deck – the lowest deck in the ship, below the waterline – where he would be better protected from cannon fire and able to operate and treat casualties relatively uninterrupted by deadly munitions.

Like all navy surgeons, Archibald had to purchase his own medical instruments and medicines. These were supplied by the Navy Stock Company, an offshoot of the London Society of Apothecaries. The cost, £ 18–25, was considerable – roughly equivalent to a full year's pay. But this would supply him with everything he would need to perform routine services, such as dental work, or more complicated procedures such as those requiring catheters for draining fluid, or trephining tools to treat a compressed skull fracture, or probes and extracting forceps for removing splinters, shrapnel, and bullets, and a variety of saws and tourniquets for amputations. In accordance with the *Regulations and Instructions*, his medical chest was examined by the medical board then locked and sealed with the marks of the Physician and the Surgeon's Company until such time as the ship set sail. This procedure assured the captain that the medical kit was sufficient and complete.

After three months' uneventful patrol with the Channel Fleet the *Nonsuch* was eventually directed to join the Caribbean fleet at the Leeward Islands under command of Admiral Sir George Rodney, arriving in January 1782.

The Royal Navy had a force of 40 ships gathering in the Caribbean, 36 of them eventually meeting up at St. Lucia. The waters here were well south of the American east coast, and the American War of Independence – which Archibald had followed from Edinburgh – was all but over by this point. But terms of peace had yet to be settled and the colonies' Spanish, Dutch, and French allies were as eager as ever to engage the British and disposes her of any possessions or influence in this region. A large force was essential to protect Jamaica, Barbados, Tobago, Saint Vincent, Grenada, Dominica, St Kitts, Antigua, and Nevis from hostile takeover by the French.

Archibald would have been excited to cross the Atlantic for the first time and keen to experience the tropical waters that in recent years had been the source of so much commerce, conversation, and conflict. And of course there would be new opportunities to discover and document exotic plants, birds, and other things of interest for his friends in the scientific community.

The Caribbean is not generally remembered as a theatre of the American Revolutionary War. Contemporary American histories typically focus on the land war and such places as Lexington and Concord and soldiers like Paul Revere. But much of the war was conducted at sea and its outcome determined by naval power. Even George Washington recognised this reality, writing to his French ally the Marquis de Lafayette in November of 1781 that 'Without a decisive naval force we can do nothing definitive'.[42] So the French navy, one of the greatest in the world, took on the British navy, the other great naval power in the world, to fight the American revolution at sea on their behalf. Between them, the two navies deployed more than 1,200 warships, armed with 25,000 cannons manned by more than 300,000 sailors in a conflict that spanned the globe from the Chesapeake to the Caribbean and beyond. The power of the naval forces deployed by Britain and France was staggering: one ship of the line could concentrate more firepower than General George Washington's entire Continental Army, and the British navy deployed more than one hundred of these. The French fleet commanded by admiral de Grasse at the Battle of the Chesapeake could deliver more than 20 times the firepower of the combined armies of Washington and Rochambeau.[43]

Joining the fleet in the Caribbean, Archibald was likely pleased to learn that Scottish surgeon Gilbert Blane was there too. Blane, who was serving as surgeon on the flagship, HMS *Formidable*, was just five years older than Archibald, and had also studied medicine at the university in Edinburgh. He had joined the navy at age thirty, and was initially engaged by Rodney to be his personal physician but was swiftly promoted to the position of physician to the fleet. By the time Archibald arrived in the Caribbean, Blane had been in that role for three years and had seen action against the French five times. In 1781 he had published *A Short Account of the Most Effectual Means of preserving the Health of Seamen*, and while it had been sent to all surgeons in the fleet the main appeal was to the commanders and officers who held overall responsibility for the health of the lives entrusted to them. To these officers he wrote:

> The health of ships companies depends in a great measure upon circumstances within the power of officers, and upon them much more than the medical branch, the health of the men depends, in as much as prevention is better than cure, and the art of physic is at best but fallible.[44]

It is a credit to fleet commander Admiral Rodney that such a progressive view towards seamen's health was welcomed and encouraged. The sixty-three-year-old Rodney was himself in excellent physical condition, which was something of a miracle considering he had by this time spent 40 years at sea, had served on over a dozen ships, and survived a life which exposed him in equal measure to armed conflict and the rudimentary sanitation and medical care of the age. His flagship, *Formidable,* had 90 cannons spread out over three decks and was the most powerful ship of the line. Fresh from action in Gibraltar, she joined the other ships of the fleet on 19 February, just a few weeks after *Nonsuch* arrived on the scene.

Rodney used the next few weeks to meet with the other commanders in the fleet and review their mission and the plans for their patrols. Blane had time to meet with them too, and with Archibald and other ship's surgeons, to review their own specific mission regarding crew health and welfare. But, after being in the Caribbean for six weeks, matters of tropical health and preventive medicine gave way to something more extreme and urgent: the French fleet had been spotted at Martinique, roughly 50 km north of the British fleet gathered at St Lucia.

The Battle of the Saintes

On the morning of 8 April, British frigates stationed with a view of the French fleet signalled to Rodney that the enemy was sailing out of their position and all 33 of their vessels were heading north. Within two hours Rodney had the British fleet's anchors up and they were all sailing in hot pursuit.

By morning of the next day Rodney's fleet of 36 ships had closed to within two or three gunshots of their target, now sailing at the north end of Dominica. Rodney gave the order to engage and the vanguard of his fleet – the front section, closest to the enemy – began a cannonade that lasted four hours and succeeded in harassing the French ships but failed to draw them into a full-fledged battle.

The British ships were not as technologically advanced as those manned by the French, and generally not as fast, but their artillery had recently been upgraded thanks to innovations by Scottish-born naval officer Sir Charles Douglas, sailing with Rodney aboard the *Formidable.*[45] Under his direction, gun mountings and ports had been improved to present a larger angle of fire both forward and back. Most of the British ships had also been equipped with mounted light, wide-bore, short-range weapons. These were formally called 'carronacks' but informally referred to by their effect: 'smashers'.

During the action on the morning of the 9th Captain Bayne of HMS *Alfred* was a fatal casualty: his leg was carried off at mid-thigh by a chain shot – a length of chain, fired by cannon, designed to rip sails and rigging apart, but equally capable of ripping people apart if they got in the way.

The British fleet continued its attempt to force the French to fully engage that day. Rodney was certain that if he could manoeuvre them into a full-on engagement – where the ships of each fleet lline up and pass by each other, exchanging blows – Douglas's innovations would allow the British ships to keep the French ships under fire for a longer period of time and win the day. So, for the rest of the day, and the one after that, the two fleets played cat and mouse. The French admiral, de Grasse, successfully directed his nimble and fast ships to deny Rodney the full engagement he desired.

For two days Archibald and surgeons on all the other ships in the fleet watched and waited on high alert for any event that would bring the frustrating chase to a climax.

Finally, on the night of the 10th, something happened that changed the game: one of the French ships, the *Zélé*, collided with de Grasse's much larger flagship, *Ville de Paris,* damaging its rigging sufficiently to make it a lame duck vulnerable to attack. The next morning de Grasse valiantly, though perhaps in hindsight foolishly, decided not to let the crippled *Zélé* fall into British hands. Confident in the superiority of his ships, damage notwithstanding, he slowed his pace in order to protect the slower *Zélé*. But this allowed Rodney, after three days of pursuit, to close in.

By the morning of the 12th the die was cast, and all the conditions Rodney wanted for battle had finally been achieved. The wind was steady, the sea fair, the sky clear, and as Blane would later write, 'If superior beings make a sport of the quarrels of mortals, they could not have chosen a better theatre for this grand and magnificent exhibition, nor could they have better entertainment than this day afforded'.[46] At half past seven, as the sun rose over the otherwise peaceful Caribbean water, Rodney gave the signal to start a battle that would last until sunset. The two fleets lined up, the British facing the French and vice versa. As the two lead ships came bow-to-bow the battle began in earnest. The immediate effect was 'one peal of thunder and blaze of fire from one end of the line to the other'.[47] Cannon balls began their relentless cascade, pounding and puncturing the oak timbers of each ship, tearing through anything in their way and blasting wooden shrapnel among the seamen above and below decks.

Archibald's ship, *Nonsuch*, was fourth in line of attack. When his ship drew parallel to the lead ship of the French fleet, the 74gun *Hercule*, the three British ships ahead of her – *Marlborough*, *Arrogant*, and *Alcide* – had each had a chance to pound *Hercule* with salvos from their guns and soften her up. That was the good news. The bad news was that *Hercule* had 74 guns of her own, 37 of them pointed towards *Nonsuch*'s starboard side, and they were already warmed up and ready to unleash their fury. The *Nonsuch* and *Hercule* exchanged fire, each blasting the other from stern to bow, firing and reloading as quickly as possible for maximum effect. The wind was firm but light, and with each vessel travelling at not much more than two miles per hour they were able to exchange fire, blow-for-blow. There was a brief pause after the two ships passed, and then it began again. Now *Nonsuch* passed the second ship in the French line, the *Souverain*. Again, thunderous cannon fire was exchanged. Each ship ripping the other to shreds in a gauntlet of fire, steel, blood, and destruction, cannon balls ripping men and rigging apart with equal indiscretion. Then another pause as they passed. Then it repeated again as *Nonsuch* passed and exchanged deadly blows with *Palmier, Northumberland*, and *Neptune*.

If, in between salvos, Archibald had left the relative safely of the orlop deck to quickly take in the scene above he would have been astounded by the size and ferocity of the sixth French ship in line, the *August*. Not only was she an 80-gun ship, she was under command of Rear Admiral Louis-Antoine, Comte de Bougainville. He was renowned for his fierce combat experience, first as an army officer defending France's North American colony of New France, and more recently as a naval commander present at the Battle of Chesapeake. But he was also known to Archibald as the first Frenchman to circumnavigate the globe. Bougainville had taken with him as botanist, Philibert Commerçon, who named the Bougainvillea flower after him. Years earlier the rear admiral had also been made a member of the Royal Society in London for his academic treatise on integral calculus. As a keen mathematician, Archibald may have imagined that one day he might meet Bougainville – though not under these circumstances.

But this was war. So, as the bow of the *August* drew parallel to *Nonsuch* their gunnery commanders each ordered their crew to fire as they passed. Archibald would have been at his station below as Captain Truscott above yelled encouragement to the crew through the gun smoke, the flying splinters of wood, the musket fire, and the roar of cannons ringing in his ears.

On shore, people could see the smoke of cannon fire roll over the water like a fog of death. The whole western half of the basin was smothered in clouds of gun smoke. On the orlop decks, each ship's surgeon attended to the wounded. On the decks above, the shattered corpses of the dead were thrown overboard to keep the decks clear of blood and to give those alive room to manoeuvre.

After two hours relentlessly giving and receiving iron-fisted body blows, the ten-mile-long British column of ships reached the point where their leading ship was alongside the rearmost ship on the French side. The two fleets' flagships, in the middle of their line, were about to pass each other. Admiral Rodney, on board the *Formidable*, looked towards Admiral de Grasse on the *Ville de Paris* and each man steeled himself for what could be the last minutes of his life. The *Formidable*, true to its name, had already taken and returned fire with 16 of the French vanguard, and had 45 guns ready to fire. But the *Ville de Paris* was even more formidable. Reputed to be the largest and most powerful ship in the world, the 104-gun, three-decked sea monster had been a gift to the state by the municipality of Paris at considerable cost. She was more than a ship; she was the pride of France and could be counted upon to deliver the full might of her resources. Across her bow the confidence of a nation was reflected in her gold-painted motto: *Fluctuat nec mergitur* (tossed by the waves, but does not sink).

Suddenly, fate intervened again as the wind, and with it the future, briefly changed direction. The *Ville de Paris* responded with a course correction that created a gap between her and the ship ahead; Rodney saw his opening and Douglas pressed him to take it. Rodney quickly directed his ship hard to starboard and broke the French line 'by bursting through it, going within short pistol-shot of the last enemy's ship we passed'.[48] Immediately after cutting the French line, Rodney signalled again to his vanguard – which included the *Nonsuch* – to tack around behind the broken French line and engage at will. The *Nonsuch* did as directed, and Archibald had to hang on to his tools and his patients as the ship turned hard to starboard, righted herself, and then let loose with cannon fire striking deadly blows from both port and starboard as targets of opportunity presented themselves, coordinating with other British ships to focus their fire – both cannon and the smashers – with devastating effect.

In a moment between the action on the upper decks, Archibald may have watched again in both in awe and horror as the 74-gun *Glorieux* was systematically stripped of all her masts and rigging and her commander, the Vicomte D'Escars, blown apart by a cannon ball directly through his chest. Breaking the line had enabled the British ships to manoeuvre behind the French ships and rake fire through their stern – the undefended and weaker rear end of the ships – sending cannon balls down the full length of the deck taking out sailors, gunners, soldiers by the score and rendering enemy ships utterly defenceless.

One after another the French ships struck their colours – signalling their surrender – until eventually even the flagship, the glorious and once-magnificent *Ville de Paris*, did the same. Blane would later write that 'The thrill of ecstasy that penetrated every British bosom in the triumphant moment of her surrender is not to be described'. As the canons fell silent and cooled, Rodney manoeuvred *Formidable* next to *Ville de Paris* so he could board her and formally accept de Grasse's sword in surrender. Blane accompanied him and recorded the scene: the deck was 'still covered with dead and wounded, [and] only de Grasse himself remained still standing, together with two or three other persons.[49] The ship

> ... presented a scene of complete horror. The numbers killed were so great, that the surviving, either from want of leisure, or through dismay, had not thrown the bodies of the killed overboard; so the decks were covered with the blood and mangled limbs of the dead, as well as the wounded and dying, now forlorn and helpless in their sufferings.

Back on the *Nonsuch*, Archibald was undoubtedly exhausted, his white shirt pink from a grim mix of blood and sweat. Going to the upper deck to gasp some fresh air and reflect on the day's events he would have seen all around him the aftermath of conflict – over 60 battleships in various states of disrepair. Not far away he may have noticed smoke, then flames, growing from the crippled French ship *Caesar*. Soon a small crowd gathered with him to watch as the flames increased in size. Then, just as they were wondering if the ship would survive the fire, and if any on board could be saved, the fire reached the powder magazine and the entire vessel exploded into a million pieces of lumber, rope, sail, and bodies – the percussion filling the sails of the ships nearby and rocking ships within its wake. On *Formidable* Blane recorded the scene:

> The French Captain, who had been severely wounded, the English officer who boarded her, together with the greater part of the men on board, both British and French, perished. Some saved themselves before the explosion; others, who survived it, and clung to parts of the wreck, were most of them either overwhelmed in the waves, or miserably scorched by the flames; and those who attempted to save them, relate that they saw a spectacle almost too horrid to mention, the men who clung to the wreck torn off by the voracious sharks, which swarm in these seas after an engagement, and were not yet glutted with the carnage of the preceding day.[50]

The historic conflict, which would be known to history as the Battle of the Saintes – after the nearby Saintes Islands – had seen all British ships equally engaged and none totally disabled. Blane gathered reports from each ship's surgeon and tallied 261 killed (including just three on *Nonsuch*) and 837 wounded. The French, on the other hand, tallied about 3,000 dead or wounded and another 5,000 captured. The human loss for France was substantial, and so was the loss of treasure: all the surviving French vessels were now property of the British navy, and 26 chests of gold and silver aboard their flagship were confiscated as well.

The battle was too late to affect the outcome of the American War of Independence in the same way the Battle of Chesapeake had done seven months earlier, but it helped accelerate its formal conclusion.

The significance of the Saintes battle was that Britain was now able to end the war as the dominant naval power on the east coast of America and in the Caribbean. France had paid a dear price for her support of the rebels – she was now lacking in money and much of her naval force was now either in British hands or at the bottom of the sea. The American colonies may have gained their independence with the help of French treasure and her naval power, but the French were now financially crushed and no longer a threat at sea.

Naval historians also point to the Battle of the Saintes as an important milestone in naval war tactics. Previously, as at engagements like the Battle of Chesapeake, the French navy had been successful against the British because it was generally faster, and was able to thwart the British strategy of lining their ships up in a row and firing on their enemy in that fashion – rather like jousting at sea. That changed at the Battle of Saintes when Rodney 'broke the line'. It was a tactic recommended by Scottish merchant, and former medical student at the University of Edinburgh, John Clerk of Eldon in his 1779 work *Essay on Naval Tactics* and one that would be repeated in 20 years by Horatio Nelson.

Saintes was also a defining personal moment for Archibald. His abilities as a surgeon and medic had been tested under fire and he had performed brilliantly. *Nonsuch* had been as engaged as any other ship but, thanks to Archibald, their fatalities were 50% lower than the average for the fleet. If anyone had doubted whether the articulate, plant-collecting, Latin-writing, Philosophy Society member was cut out for a career at sea, they had no doubts now. The battle was also an experience that would forge a lifelong bond of comradeship between him and people he would meet in the weeks and years ahead; for nothing so binds one man to another as the shock, awe, horror, and shared experience of war. Present this day among the fire and thunder were: Peter Puget, a seventeen-year-old midshipman serving on HMS *Alert*, under command of Captain James Vashon; fifteen-year-old Joseph Baker, also on *Alert*, serving as Vashon's cabin boy; and James Johnstone, a twenty-three-year-old lieutenant, on *Formidable*. Fourth lieutenant George Vancouver missed out on the Battle of the Saintes, arriving shortly after it was over, but joined HMS *Fame* in Jamaica and shared in the fleet's claim to glory. He would later serve under Vashon aboard HMS *Europa*.[51] Though none of them could know they would one day serve together on the *Discovery*, the bonds between them were established here in the Caribbean.

News of their victory arrived in London about a month later, on 18 May, and immediately electrified the city. One of Rodney's daughters wrote to him, describing the reaction there:

> London was in uproar; the whole town was illuminated at night. We were at the play. When we went in, the whole house testified, by their claps and huzzahs, the joy they felt at the news and their love for you. Their acclamations lasted for, I am sure, five minutes.[52]

For the next few months the fleet stayed in and around Jamaica, basking in the sun and in their glory. Archibald's competent combat service now earned him the honour of serving on

Formidable, and his evident skill and keen intellect made him a useful and desirable member of Blane's medical staff. He also developed a friendship with the like-minded James Johnstone, and when the fleet put in to Port Royal for recovery and repairs, the two likely enjoyed touring what had once been the notorious home port for pirates, privateers, and various rogues who had preyed upon Spanish ships laden with gold and other treasure. They probably also toured the nearby town of Kingstown, whose population of about 11,000 were mostly engaged in the incredibly profitable and strategically important sugar trade. Here they were also confronted with the stark reality of the slave trade, which provided the manpower for the plantations even as it undermined the morals and humanity of an empire.[53]

In July Admiral Hugh Pigot arrived to take over command of the fleet while Rodney returned home to glory. The *Formidable* and her fleet patrolled the area for the rest of July and August, but by September events were such that a naval presence was needed further north to support the safe evacuation of those American colonists who had supported the Crown and were eager, or otherwise compelled, to leave.

Nova Scotia

By 5 Sept 1782 the *Formidable* arrived at Sandy Hook, the 9 km-long peninsula that guarded the entrance to New York harbour just 12 km further north, using the lighthouse there, built in 1764 and in British control since 1776, to help them navigate safely to port. Here, Archibald transferred to HMS *Assistance* for further transport to his next posting at the navy base at Halifax, Nova Scotia, 900 km to the northeast. The *Assistance* was not nearly so grand as *Formidable* and, at 44 metres long with just 50 guns and a crew of only 350, was even smaller than *Nonsuch*. But she was a fine ship all the same, and he would not be alone: James Johnstone also transferred from *Formidable* to *Assistance*, and both of them would be under command of Saintes battle veteran Charles Douglas, now appointed commodore and commander-in-chief of Halifax Station.

The city of Halifax at this time had a population of a just a few thousand people, but the naval yard to the north was 'completely built with supplies and stores of every kind for the royal navy. The harbour of Halifax is reckoned inferior to no place in British America for the seat of government, being open and accessible at all seasons of the year when almost all harbours in these provinces are locked up with ice.'[54] After his arrival there, Archibald relayed his impressions to his mentor and patron in Edinburgh, John Hope, writing,

> I arrived here from the W Indies in the Assistance of 50 guns a few days ago, and I am charmed with the general appearance of the country which seems to offer a most delightful prospect for Botanical researches, as well as other branches of Natural History. I already had two excursions into the woods and I cannot describe to you the pleasure I felt when surrounded with the *Kalmia angustifolia, Andromeda calycantha, Ledum palustre, Gaultheria procumbens, Arbutus uva-ursi, Pinus strobus, P canadense* and several other beautiful evergreens which I could not ascertain, besides a vast number of Cryptogamia plants of which you know I am passionately fond.[55]

Still nurturing his dream of further and more exotic adventure such as Banks had enjoyed with Cook, he promised to keep working on botanical pursuits in his spare time: 'I shall wholly devote my vacant hours to natural history while I remain on this station and I have no doubt that but I shall be able to send you another parcel of seeds and specimens in the autumn'.

Archibald was wise to get out of town while and when he could, as Halifax was soon to become a very busy place and not entirely unchaotic. The peace negotiations between Great Britain and the American revolutionaries, drafted in November of the previous year, reached a milestone in January of 1783 and some 30,000 of the 50,000 loyalists who had already fled to New York from the more southerly states were now eyeing Halifax as a refuge from persecution where they might also shape a prosperous future. The negotiations had conceded American access to the inshore fisheries of British America under the Treaty of Paris, but the London authorities also introduced imperial Orders-in-Council barring American vessels from British Caribbean ports – which had been secured thanks to the successful action at the Battle of the Saintes. Halifax entrepreneurs had every reason therefore to expect their small town could soon become as significant as Boston, and in time be a major regional commercial metropolis, thriving on the Caribbean trade and functioning as chief commercial access port within the remaining and not inconsequential British interests in the region.

Among the thousands of refugees to Halifax were many former slaves. This was due in no small measure to a formal declaration made during the war by Virginia governor Lord Dunmore offering freedom to any slave who took up arms with the British forces against the revolutionaries. Nearly 100,000 enslaved people had taken advantage of that offer, betting their lives and liberty on a British victory, but were now at risk and desperate to settle in British-controlled territory.[56]

Archibald would have been startled to learn that one of Halifax's new residents in the summer of 1783 was an escaped slave from General George Washington's Mount Vernon home in Virginia: Harry Washington, having made his way to New York, had managed to get passage on the HMS *L'Abondance*, a British ship which had been captured from the French a few years earlier. When *L'Abondance* arrived at Halifax Station, docking near *Assistance*, Archibald would have learned from local chatter that the forty-three-year-old Harry Washington had joined Dunmore's all-black British regiment, proudly wearing a uniform embroidered with the motto 'Liberty to Slaves'.[57] Washington was one of a few thousand former slaves who made their way to Halifax that summer, joining some 1,500 others who had settled in Birchtown, on its way to becoming the largest free Black community in North America. It was a stark contrast to the scenes Archibald would have observed in the port town of Kingston Jamaica not too long ago.

Being attached to the military base at Halifax, Archibald would also have met the many other sons of Scotland there. The 84th Royal Highland Regiment had been stationed there since the fall of 1782 and were well-established by the time Archibald arrived. The 42nd Regiment of Foot – now known as the Black Watch – also all highlanders, arrived later, in January 1784. Archibald enjoyed the company of the many hundreds of fellow Scots in the area but was particularly pleased to strike up a friendship with Colonel John Small of the 84th. It turned out that the fifty-seven-year-old had been born just 25 km west of Archibald's home near Aberfeldy, and they had much in common. In Halifax and at Small's country manor in Hants County they shared memories of the Highlands and compared notes on the geographic similarities between Scotland and Nova Scotia. Small shared how his older brother was also a surgeon – in the army – who had presented a medical paper in Edinburgh while Archibald was a student there. Archibald wrote to Hope in

1785 saying that Small had even invited him to spend part of the next summer with him at his country seat near Windsor, that is, if they didn't go to Quebec instead.[58]

Small's 84th Royal Highland Regiment comprised some 2,000 men formed entirely of Scottish soldiers who had served in the Seven Years War and the American War of Independence and, as such, composed some of the most seasoned and experienced officers and men in North America, and Sharp had many excellent stories for Archibald. The seasoned veteran had served in the Netherlands with the Scots Brigade before joining the 42nd Regiment of Foot – which Archibald knew had first mustered in Aberfeldy in 1740 – just in time to join their action in Quebec. He had a perspective and experience on the land, the various Indigenous peoples, the Europeans, and the politics, which Archibald would have found unique and stimulating. Small had seen a great deal of action but his story about his experience at the Battle of Bunker Hill was particularly riveting. It was a narrative he would repeat again to artist John Trumbull when he was back in London the next year, and when Trumbull completed his epic painting *The Death of General Warren at the Battle of Bunker Hill*, he would actually place Small at the centre of the action, preventing a fellow British soldier from bayonetting the tragic hero of the scene, Joseph Warren.

Small may have confided to Archibald that Warren had actually been cut down by a musket ball through the head, but artist Trumbull was so touched by the humanity and kindness Small had shown to his enemies that he thought this a more poignant scene.

While all that activity was transpiring on the land, the sea lanes were occupied with traffic as both military and commercial ships busily transported people and goods from New York to Halifax, and Douglas was knee-deep in the business of interpreting and applying treaty obligations with the new United States. Archibald continued his duties as surgeon on *Assistance* but also made good on his promise to Hope. Over the next three years he would use his 'vacant hours' to collect, document, and send a host of botanical observations, seeds, and samples to Hope, Banks, Pitcairn, Fothergill, and others in the natural science community back home.

To Anna Blackburne, one of only three female naturalists known to Carl Linnaeus, Archibald sent material to whet her interest in North American birds. He toured the western part of Nova Scotia, examining bogs, marshes, and small lakes; he scoured the shores of St John's Island (now Prince Edward Island), returned to Sandy Hook and New York to gather samples during trips there; and did the same from voyages back down to Barbados, Dominica, St Christopher's (Saint Kitts), and Nevis. With him, he carried his notebooks and over his shoulder an oblong tin canister – called a vasculum – to carry freshly-collected plant specimens and keep them fresh until they could be properly pressed.

In the ships sailing to and fro across the Atlantic, Archibald received from London four volumes of the 12th edition of Linnaeus' *Systema Naturae* and the supplement to the 13th edition of Linnaeus' *Systema Vegetabilium*, and from Hope he also received copies of his short *Genera* and a catalogue of trees and shrubs. He was thousands of miles away from Edinburgh and London, but nurtured and cultivated his relationships with John Hope and Joseph Banks through steady correspondence and by using his intellect, personality, and physical stamina to retain a presence in their minds.

In the six years since leaving Edinburgh, Archibald had experienced both blood and botany, and he was keen to experience more. He supposed the way to achieve this was to leverage his skills in botany so he could be assigned to an exotic voyage of discovery; this would give him they key to following in the footsteps of Cook, Banks, Solander, and the other great gentlemen of action and adventure who fired his imagination and fuelled his ambition.

He would soon get the chance to put that plan into action.

Four
SOHO SQUARE TO NOOTKA SOUND (1786–87)

While still sailing around the east coast of the American continent, Archibald heard through his network of contacts that a trade mission was being planned for the relatively uncharted, unbotanised, unexplored Pacific west coast of North America. It was the break he was looking for, and he appealed directly to Joseph Banks for a chance to join it.

Sweetening his pitch with the prospect of gathering more seeds, he wrote to Banks, 'I am informed that there is a Ship, a private adventurer now fitting out at Deptford to go round the world. Should I be so happy as to be appointed surgeon of her, it will at least gratify one of my greatest earthly ambitions, and afford one of the best opportunities of collecting Seeds and other objects of Natural History for you and the rest of my friends … '[59]

The mission Archibald wished to join was being managed by the enterprising London tea and wine merchant Richard Etches. Etches had read the account of Cook's third voyage, published by the Admiralty in 1784, and was intrigued by the potential for trading sea otter pelts from the Pacific Northwest to China. He was excited to learn that profits from such trade could be huge – certainly high enough to justify the financial risk of putting a trade mission together. It was said that traders in Macao, China, would pay about 120 Spanish dollars (about thirty British pounds – roughly US$ 2,000 in current value) for just one skin. A cargo of 1,000 top quality skins might be worth a couple of million US dollars today.[60]

Archibald was wise to write to Banks, and his timing was perfect. Etches had been courting Banks's support for some time and was motivated to please him. He needed Banks's help because the markets at Canton lay within the territory of the East India Company's trading monopoly and, to profit from the fur opportunity legally, Etches needed their permission to trade there. With Banks's help, he had received a five-year licence[61] and had already sent two ships: the 320-ton *King George* under command of Nathaniel Portlock and the 200-ton *Queen Charlotte* (so named by Joseph Banks) under command of George Dixon.

The two captains were both in their mid-thirties but Portlock had thirteen years' experience in the Royal Navy compared to Dixon's nine and was the more senior of the two both in age and rank. They were also both veterans of Cook's third voyage and had been with him to Nootka, as well as to Hawaii and to the great trading port of Canton. They stood to make significant gains in their personal fortunes as partners in Etches's newly formed King George's Sound Company, which naturally had Richard Etches as its principal founder and included six other investors: Richard's brother John Etches; two of his relatives, both named William Etches; a gentleman from

Devon named John Hanning; a merchant from Hampshire named Nathaniel Gilmour; and a woman tea dealer in London, Mary Camilla Brook.

When Banks received Archibald's letter, he knew Etches was already organising a second expedition to sail shortly after Archibald was due back from Halifax. Thirty-three-year-old James Colnett, who had sailed with Cook on his second voyage as a teenager, had already been identified at the mission commander. But other roles had yet to be determined.

As soon as Archibald arrived back in London in August, 1786, he immediately introduced himself to Banks, supported by a reference letter from Professor Hope. In his letter Hope assured Banks that during Archibald's recent adventures in the Caribbean and around Halifax he had 'paid unremitting attention to his favourite study of botany, and through the indulgence of the Commander-in-Chief had good opportunities afforded him'[62]. Banks of course already knew Archibald indirectly through his correspondence, and by reputation through mutual colleagues like Fothergill and Pitcairn, and his impressions were positive. Nevertheless, their first face-to-face meeting was important. It was perhaps the most important interview of Archibald's life – and fortunately he passed muster: Banks approved of his character, knowledge, and skill, and promptly wrote to Etches suggesting he add a scientific element to his upcoming trade mission ... and take Archibald along as surgeon and botanist.

To cement the deal, Archibald arranged to meet personally with Richard Etches to review his credentials and gain his confidence. This meeting also went well. Archibald successfully persuaded Etches that he could serve as surgeon – and botanist – without neglecting either role.

Etches subsequently wrote to Banks and explained that while his first two ships had been purely commercial the next would indeed include the scientific element Banks desired:

> Sir, I was duly honor'd with your very kind favor to Mr Menzies, to which I feel myself bound to pay every possible attention. I believe you were fully acquainted with the restrictions laid down in the former ships [*King George* and *Queen Charlotte*], in a young undertaking and of such a nature as the present I presume such restrictions absolutely necessary, but in the present instance it is my full intention to dispense with them in the case of Mr. Menzies, so far as can have any tendency to be beneficial to science in general, I highly approve of his conduct and manners, and my young brother, who is part proprietor, is going the voyage, I have gave him orders to pay every attention to Mr. M, and to give him ample latitude in his pursutes, and I have no doubt on his return, he will confess haveing experienced that liberality which your recommendation Sir most certainly demands from me.[63]

While the newly formed King George's Sound Company was putting the final details of this second mission together – selecting ships, hiring key personnel, obtaining supplies, charting routes – Archibald quickly did everything he could to prepare himself for success. Several days were likely spent at Banks's substantial London home at 32 Soho Square, which in addition to being a residence was also an impressive private library and natural history museum.[64] In his home and at his library were collections – all manner of drawings, reports, samples, artefacts, specimens – which he had personally gathered on his voyages to Newfoundland, the South Pacific, Iceland, and the Hebrides, and gathered from all his correspondents, agents, mentees, and friends. Banks's herbarium also had specimens that had been collected at Alaska and Nootka during Cook's third expedition. Banks's library may have also included the latest work by Archibald's comrade from HMS *Formidable* Gilbert Blane. The esteemed naval surgeon had just published his work *Observations on the Diseases of Seamen*.

Preparation

In addition to his research on what was known, and what remained to be known, about the natural history of the Pacific Northwest and the South Pacific areas he was soon to visit, Archibald would have consulted with others trained and experienced in this fascinating and rapidly unfolding field of knowledge. Banks's librarian, the Swedish academic and botanist Jonas Carlsson Dryander, would have been one of his contacts. Dryander was six years older than Archibald, educated at universities in Gothenburg, Lund, and Upsala, and was a former student of Carl Linnaeus.

Another person he likely connected with in preparing for his voyage is William Aiton. The fifty-five-year-old Scotsman was director of the king's gardens at Kew, just 12 km from Soho. Like Archibald, Aiton started his career as a gardener and had been an assistant to Phillip Miller, superintendent of the Chelsea Physic Garden. The two would have discussed the medicinal properties of various plants, including those he was about to encounter first-hand on this new voyage, and he would have checked on the status of the seeds Archibald had sent along previously from Halifax.

The moments spent at Banks's home, and among his associates, would have been as energising as they were educational. No place in the world had more information, more intellectual activity, more adventure seekers, and more characters, than Banks's home. The neighbourhood was impressive too. The offices of the Secretary of State were less than a five-minute walk away, as was the impressive British Museum – another source of wonder, research, and, if things went well, perhaps a future home to Archibald's collections. When visiting the museum, which had opened to the public just 25 years previously, Archibald would have visited the collections of former museum librarian Daniel Solander. The former student of Carl Linnaeus had also been a close friend of Banks and had travelled with him on Cook's first voyage, gathering, with Banks, over 30,000 specimens including 1,300 species new to Western science.

In a sombre moment of personal reflection, Banks may have related to Archibald how Solander, his close friend and scientific collaborator, had suffered a brain haemorrhage and dropped dead in Banks's herbarium just a few years earlier.[65]

As he had done in Edinburgh as a student, Archibald may have also frequented the local coffee houses and taverns to talk about matters philosophical, commercial, scientific, and political. It was said that The Turk's Head tavern on Gerrard Street had been a favourite of Scottish economist Adam Smith. In the Marlborough Coffee House, just a five-minute walk away from Banks's Soho mansion, conversations covered a range of topics from politics to news, daily gossip, fashion, current events, philosophical debates, and discussion about the natural sciences. Fellow scientists, botanists, physicians, surgeons, merchants, explorers, rubbed shoulders informally and traded news as easily as they traded opinions.

In addition to preparing his mind for the Etches mission, Archibald had to act fast to prepare his quarters and supplies aboard his next ship, the *Prince of Wales*. She was a thirty-four-year-old ship of average size (171 tons). The crew was small compared even to the *Assistance* – just thirty-five men of various occupations. About a dozen were regular seamen but the rest had speciality roles such as gunner, blacksmith, sailmaker, cook, and quartermaster.

Accommodation for the first part of the journey would be tight, since they also had to transport an associate of Richard Etches (Captain Marshal) and about 15 of his workers to the Isla de los Estados, at the southern tip of Argentina to establish a sealing outfit there.

Although the *Prince of Wales* was a trading ship she was, as a matter of precaution, armed with fourteen guns.

Arriving at the impressive Deptford Dockyard, Archibald would have been both awed and inspired by the sight of so many frigates and warships being built, or refurbished, in the dry dock and wet docks. Carts and wagons came and went with supplies and materials destined for the storehouse, the smithy, the planking shed, the rigging and sail house, and for the ships themselves. Archibald remembered that Cook's ship *Endeavour*, used on his historic first voyage, had been refitted here; and he knew here, also, Cook's ship *Resolution* had been converted from a coal-carrying merchant vessel to a proper research vessel suitable for round-the-world adventure and discovery.

When he eventually located and boarded his new home – the *Prince of Wales* – Archibald would have begun the work to ensure the surgeon's quarters were properly equipped with a medicinal kit including herbal and organic medicines in both liquid and powdered form, and a surgical kit including the usual hardware of amputating knives, an amputating saw, a metacarpal (hand bone) saw, artery forceps, tenaculums (surgical clamps), petit screw tourniquets, bone nippers and turnscrews, tooth forceps, and various splints and bandages.

The mission

The second expedition of ships from the King George's Sound Company began in earnest when the *Prince of Wales* slipped out of Deptford on 23 September 1786 and made way for the other side of the globe. Archibald's three-year voyage on the *Prince of Wales* would be unlike any other he had taken: his first around the world, first time sailing below the equator, first visit to Argentina, Nootka, Hawaii, Alaska, China, and Sumatra. And, if he survived, it would set him up for future fame, adventure, and as yet unknown fortune … or at least a wealth of scientific knowledge. He was excited beyond imagination.

To his great delight his old friend, and shipmate from *Formidable* and *Assistance*, James Johnstone was appointed as the ship's chief mate – second in command.

A second ship, the *Princess Royal* accompanied them and would provide additional support. The smaller (65 tons) tender ship was under command of Captain Charles Duncan, also in his early thirties. He too was a veteran of the American War of Independence and was also a recent shipmate of James Burney – a veteran of Cook's second and third expeditions. Among its smaller crew of 15 hands were men with roles as sailmaker, carpenter, cooper, cook, armourer, steward, and various mates. Archibald was pleased to learn that some of the crew on both ships were amateur musicians and had brought with them an organ, pipes, fife, and drum to amuse themselves while at sea and to entertain any Indigenous peoples they met on land.

Their mission was not without danger but it was simple: to acquire furs from the Indigenous peoples of the Pacific Northwest and trade them for profit in China; then use the proceeds from the sale of furs to buy other goods in China such as tea, silk, and porcelain and bring those back to London to sell at a profit. To that end, the ships had been loaded up in London with items to trade with the Indigenous peoples: quantities of iron, copper, and glass beads.

Colnett's instructions were to approach the western coast of North America at 45 degrees latitude, then proceed north, entering all the bays, harbours, sounds, creeks, rivers as they saw fit 'consistent with the true intent of a commercial voyage'.[66] Visiting Nootka Sound, where Cook had briefly established contact eight years earlier, and which Cook had identified as a strategic and opportune trading port, was an explicit objective. It was also hoped Colnett could purchase land from the Indigenous people there for construction of a trading post.

The first leg of the journey was uneventful. By November they reached the west coast of Africa (Sao Tiago), where they stopped to replenish supplies and provisions. Things became livelier on the first day of December, when the *Prince of Wales* crossed the equator and Archibald witnessed the centuries-old ceremony of shaving and 'ducking' the seamen who had never crossed the equator before. Third mate Andrew Taylor, age about twenty-five, wrote of ducking in his journal,

> This amusement not only occasions Mirth but exercise which makes the encouragement of it commendable in all Ships from their Salubrious qualities we had many on board who had never paid their devoirs to the Trident Monarch, so that the Crew had their full scope of diversion, and were Issued an allowance of their favourite grog on the merry occasions.[67]

Crossing the equator 20 years earlier with Cook, Banks provided a more detailed description of the ceremony. He said men were tied to a rope and dunked overboard from the ship's main yard:

> The Boatswain gave the command by his whistle and the man was hoisted up as high as the cross piece over his head would allow, when another signal was made and immediately the rope was let go and his own weight carried him down, he was then immediately hoisted up again and three times served in this manner which was every mans allowance. Thus ended the diversion of the day, for the ducking lasted till almost night, and sufficiently diverting it certainly was to see the different faces that were made on this occasion, some grinning and exulting in their hardiness whilst others were almost suffocated … [68]

Archibald was likely amused by these seaborne shenanigans; but as ship's surgeon he was also very much relieved that no one on the *Prince of Wales* was injured or drowned by them.

By January of 1787, Archibald and his shipmates made it to the southern end of Argentina and stopped at Isla de los Estados to drop off Captain Marshall and his party near the north side of the island.[69] They were now halfway to Nootka Sound. The two ships remained here for three weeks while Marshall and his men built their shelter and set up their commercial operation … or what Archibald called 'a Seal Fishery scheme'.

Colnett and his crew were obliged to linger and help Marshall get properly settled, since the sealing operation was another of Etches's many business interests.[70] But he was worried about wasting time and anxious to move on to Nootka. Meantime, it was a suitable place to gather fresh water, and whatever fresh edible plants could be found, to prepare for the next leg of their journey around the tip of Chile, on to Hawaii for refreshment, then finally to Nootka.

Archibald understood but did not share Colnett's anxiety. He enjoyed the layover at Isla de los Estados and found it an excellent opportunity to explore the land and conduct a survey of its natural resources. On Sunday 28 January, a day of rest, Archibald, ever the Scottish hillclimber, persuaded Colnett and Duncan to join him on a trek to the top of a nearby hill to better see the coast. Johnstone stayed behind to oversee the two ships and their crew. The highest hill on the island was only two-thirds as high as Ben Lawyers near Archibald's home in Scotland, and the one they climbed this day was an easy pursuit. The three men – two captains and their surgeon, all in their thirties – enjoyed the opportunity to ramble along the countryside, challenge their physical skills, and the two captains were entertained and inspired by Archibald's enthusiastic expositions and observations on the native plants, birds, and animals. He hoped the fresh air and physical exertion provided some relief for Colnett's pent-up energy.

Two weeks later, on 12 February, the two ships finally left Isla de los Estatdos. The original plan had been to stop in Hawaii on the way to Nootka, to take on fresh water and food, but Colnett surprised everybody when he decided to plot a course directly north after they cleared Cape Horn. He would later regret this decision, as it would almost cost them their lives. But the three weeks they spent in Isla de los Estados had worn out Colnett's patience and he was anxious to make up for lost time. He worried that other ships, working for other companies, might also be sailing to Nootka at this very moment, and he didn't want to arrive after they had bought all the best furs.

Colnett's decision was influenced by some unverified Spanish charts indicating an alternative resting place to Hawaii – the Gallegos – and he made the mistake of betting on the accuracy of those maps. In his log Colnett explained, 'I intended, could I make the Gallegos, to touch there for some refreshments, we were all at this time in perfect health … '[71]

Unfortunately, the Gallegos were elusive and by April Colnett gave up looking for them. He blamed the Spanish for misleading him:

> I now gave over all hopes of seeing it which I account for as said only to be known to the Spanish who, tenacious of their rights & privileges in these seas, ever Jealous of any navigating them but themselves, conceive it will be to their advantage to confuse an account of any discovery they make where refreshments can be had …

He concluded 'I determin'd to proceed to Nootka.'[72] He had no alternative but to stick to the course he had laid out and see his decisions through.

Colnett led the mission ever northward towards their destination. It was not naval warfare, but do or die. The life of the crew was in the balance.

For the first time on the mission Archibald had reason to be seriously worried. He knew better than anyone else the perils of prolonged absence of fresh food and, as ship's surgeon, the duty to keep everyone alive was his. He knew the risks and signs of scurvy – bleeding gums, easy bruising of the skin, poor healing of any wounds and, in extreme cases, even death – but treatment was still experimental, and his medical skills would now be tested to their limit. He knew that citrus fruits, berries, potatoes, tomatoes, peppers, cabbage, brussels sprouts, broccoli and spinach were helpful in reducing risk of scurvy but also knew that ships could only stock produce for a limited time, and an extended voyage such as this 21-week marathon by Colnett would push the crew's health to the limit.

Fortunately, Archibald was not unprepared. He had read the *Treatise of the Scurvy* by Dr James Lind and knew that current remedies included such things as a half pint of seawater a day, a quart of cider a day, two spoonfuls of vinegar three times a day, twenty-five drops of elixir of vitriol three times a day, and a nutmeg-sized paste of garlic, mustard seed, horse-radish, balsam of Peru, and gum myrrh three times a day. Lind had also conducted some interesting experiments with citrus fruits but had not definitively recommended them.

Archibald was also familiar with the more recent pamphlet published by Dr Gilbert Blane – whom he served with on *Formidable* – titled *On the most effective means for preserving the health of seamen, particularly in the Royal Navy*. Blane advocated using citrus juice to prevent and cure scurvy.[73] His patron, Banks, advised a treatment of spruce beer: a mix of spruce needles and molasses.

Despite all this information about the symptoms of scurvy and its various treatments, the exact cause and cure remained unknown. Archibald just had to do his best and fortunately he had made sure both ships were fully loaded at Isla de los Estados with fresh vegetables – such as could be found among those rocky isles – including wild celery, birch berries, and winter bark plants. These fresh plants, loaded with antiscorbutic properties, wouldn't stay fresh for ever but Archibald insisted they be included in the crew's diet. Even so, they were so afflicted with scurvy by the time they neared Nootka they barely made it at all.

Approaching Nootka, the crew were so weak that some couldn't crawl out of their bunks; anyone physically capable of manning the sails – including Archibald – had to do so, otherwise the ships risked being lost at sea. Colnett wrote a desperate note in his journal: 'The Scurvey at this time began to make its appearance among us … a Regimen of diet was prescribed by the doctor [Archibald] … All our hopes of relief from this fatal disease depended on a fair wind and quick passage'.[74]

Nootka Sound

When they did eventually come within sight of Nootka, on 5 July 1787, they were so grateful for deliverance that (Colnett observed) no one remained below who was not able to crawl up on deck to see that their ordeal was over. They had been at sea for nearly 150 days and sailed non-stop from the bottom of South America to within striking distance of Alaska. Their survival was a miracle.

As their ship limped into the safe harbour of Friendly Cove, they were delighted and relieved to see another ship was already there and ready and able to help restore the weakened crew. The substantial 400-ton[75] *Imperial Eagle*, a trade ship under command of Captain Charles Barkley, had arrived a few weeks earlier and was pleased to trade some fresh food and other supplies in exchange for some paint oil and black varnish. Colnett was grateful for the help. But after recovering his health he realised his fears had come true: Barkley had arrived in Nootka three weeks ahead of him and had already purchased all the best furs – about 700 of them. Worse, Barkley didn't have a licence from the East India Company and was essentially trading illegally. Colnett reflected later in his journal: 'I did not mention to him then the illegality of his trading in the Southsea Company's limits, thinking it would have been a breach of friendship, … he engrossed the whole trade & ruined ours'.[76]

Colnett could only wonder how much better his situation would have been if Etches had not insisted they stop in Isla do los Estados for three weeks, and how much healthier they would all be if he had not felt compelled to risk bypassing Hawaii to save time.

So much had been invested and risked, and now it all seemed to be for nothing. First, their health had declined in the act of getting here, and now their prospects of wealth also seemed to be quickly fading.

Archibald, the most capable medic within a thousand miles in any direction, left these business worries to Colnett and immediately set to work restoring the crew's health and gathering intelligence about their situation. One unexpected source of information was a Robinson Crusoe-type character by the name of John McKay. The local resident had not been shipwrecked, but he certainly looked as though he had been.

Archibald discovered that the bearded and emaciated McKay had sailed from Bombay to Nootka the previous year on a ship owned by James Charles Stuart Strange, an agent of the East India Company.[77] Like the Etches brothers and their enterprising business associates, Strange was keen to take advantage of the fur trading opportunities identified by Cook. McKay told Archibald that Strange had financed two ships to sail to Nootka the previous year to trade for furs and bring them back to China for profit: the 350-ton *Captain Cook*, and the 100-ton *Experiment*. McKay was on the *Captain Cook* as surgeon's mate but took ill just before it was about to return to China and volunteered to stay behind. The Nuu-Chah-Nulth Chief Maquinna said he could remain as his guest so that he could recover his health and learn the language, customs, and ways of the local people. Learning the local language and their ways would help advance the opportunities for further trade. Strange told McKay he would arrange for him to be picked up again the next year.

The plan was successful for the first few months; but things took a turn for the worse when McKay – not yet fully cognizant of local customs – broke a taboo by stepping over the cradle of one of Chief Maquinna's children. That earned him a beating. But, when the child died some time afterwards, McKay was lucky not to receive harsher punishment. Instead, he was removed from the Maquinna home, ostracised, and spent the next year living on the edge of society, fending largely for himself.

Although McKay was in rough shape and had enjoyed only limited contact with the local Indigenous tribes – his self-prescribed cultural immersion programme falling far below expectations – he was nevertheless the first white man to spend a full year in the area and consequently the most knowledgeable English-speaking source of information about the language, customs, and traits of the local people. Archibald interrogated him thoroughly and after many animated and probing discussions with McKay, learned the important role local women played as caretakers and physicians within the tribe. He then began to seek those women out so he could learn about their ways of knowing and perhaps discover new plants or new medical treatments, in addition to sharing what other knowledge he had about physic gardens.

McKay helped explain to the local women that Archibald was an expert in plants and medicine and facilitated some understanding between them before he gratefully left with Barkley.

Older, wiser, gentler, more inquisitive, better educated, and better prepared than any previous European visitor to Nootka, Archibald would have clearly stood out from the other Europeans, and it was not difficult for him to quickly earn the trust and respect of the women in the area. His interest in plants and knowledge of their medicinal and other properties set him apart from all the men who had visited before. The daughter of an elderly chief especially took him into her care and made sure he was watched over. Archibald later wrote to Banks explaining the women

... often warned me in the most earnest manner of the dangers to which my Botanical rambles in the Woods exposed me & when they found me inattentive to their entreaties, they would then watch the avenue of the Forest where I entered, to prevent my receiving any insult or ill-usage from their Countrymen.[78]

To his utter amazement, Archibald also met a Hawaiian woman at Friendly Cove. Wynee was sailing on Barkley's ship, having joined the *Imperial Eagle* when Barkley stopped in Hawaii some weeks earlier.[79] Joining the voyage had apparently been her idea, but Barkley described her as his maid servant. Archibald was likely thrilled to talk with her and to learn some of her language and ways, just as he had been doing with the chief's daughter and her friends. Barkley astonished everyone further by introducing them to his seventeen-year-old bride Francis. She had married Charles just nine months earlier.[80]

Archibald's conversations with Francis, Wynee, and the chief's daughter and her friends helped accelerate his understanding of the culture he had just encountered, and the others he soon would.

Charles Barkley, though a commercial competitor and operating rather unfairly without a licence from the East India Company, was an interesting man in his own right. He was just four years younger than Archibald and descended from a well-connected line of Barkleys from Aberdeen, Scotland who had fallen on hard times after the 1746 Battle of Culloden. It would not be surprising if Archibald and Charles traded stories of their youth, and the events that had led them to this place and this unexpected but fruitful meeting. If they did, Archibald would have learned that Barkley's life in the merchant navy began at the age of eleven, sailing on the East India Company ship *Pacific*, and that his sea-faring experiences also included some time sailing in the Caribbean. Archibald may have entertained him with stories of his time in those seas, and especially his recent experiences confronting the massive battleships of the French and Spanish. The conversations may have reminded them both of days at the London Coffee Houses – Barkley explaining that the St Paul's Coffee House near the famous Cathedral served as his home address when there.

In addition to these informal, friendly conversations, Archibald was probably able to gain some practical insights from Barkley regarding his past few weeks trading in the area and engaging with the local people, learning aspects of their language and customs. He likely learned from Barkley that Chief Maquinna and his brother Callecum were important and influential figures in the region.[81]

Meanwhile, following some recuperative time under Archibald's medical care and attention, Colnett's log reported on their gloomy business situation:

The few skins & pieces we had collected from the Natives for the length of time we had been here, and the Natives informing me, Captain Berkeley of the *Lowden* (the previous name of the *Imperial Eagle*) had purchased them all, gave me little hopes of being able to make any returns to my Owners; the ship's crew being mostly able to do something we began to wood, water, & fit out with every expedition to proceed to the Northward, in hopes of better success.[82]

Having seen to the health of the crew, Archibald did his best to pitch in and further the trading interests of the mission. Sailing around the area he soon discovered that nobody was much interested in the trading beads they brought with them.

He reported this insight to Banks, for future reference:

> At Nootka we found copper the article most sought after and in this we were deficient, having little or none aboard. At Prince William's Sound the natives preferred iron and put very little value of anything else – they were so overstocked with beads as to ornament their dogs with them.

Speaking of the peoples at Queen Charlotte's Isles (Haida Gwaii) and Banks Island, he added that 'iron, cloth beads with brass and copper trinkets answered best' while at Cape Edgecombe 'iron frying pans – tin kettles – pewter basons and beads formed the chief articles of trade. Ornamental lofty caps with brass or copper would be good presents for the chiefs and warriors'.[83]

Archibald, Captains Colnett and Duncan, and chief mate Johnstone were all new to the region but they knew they were not the first to trade there. The Indigenous people they encountered at Nootka and elsewhere in nearby waters had been trading for generations. Abalone shells had made their way here from as far south as California, traded up the coast over months and years.[84] Copper nuggets occurred and were found naturally, while small quantities of iron had arrived, from time to time, via driftwood wreckage from Asia. Metal was rare and highly prized, but not unknown. More of it had passed through their hands in recent years than ever before, as first the Spaniard explorer Pérez visited (though did not land) in 1774, followed by Cook in 1778, then fur traders Strange and Hanna in 1786, and now Barkley, Colnett, and Duncan in 1787. Nootka, and specifically the village at Friendly Cove under authority of Chief Maquinna, was poised to become a major port connecting global trade to the north-west coast of America.

Thinking of ways to improve their trade relationship, Archibald made a note to tell Banks that future expeditions to the area should not bring iron and copper alone, but include blacksmiths and a forge, so that metal bars and ingots could be fashioned to suit local needs and desires.

Archibald's journey from Soho Square to Nootka Sound had been long and at times perilous. The fate of the voyage seemed now more about recovery and survival than profit. But for Archibald, whose motivation was adventure and discovery, the voyage was delivering all the riches he had hoped for.

He could only hope the remaining two years of his voyage would be as rewarding as the first.

Five
COLLECTING, AND CONNECTING (1787–1789)

Once they had recovered their health, and their bearings, the crew of the *Prince of Wales* and *Princess Royal* spent the next five months (July to November 1787) sailing around the islands and inlets between Nootka and Alaska, searching for fur and charting the coastline. At each opportunity Archibald eagerly collected botanical specimens and made personal connections with the different Indigenous tribes, trading for artefacts and information.

Every day was a day of discovery, and Archibald, the first university-trained scientist to visit many parts of the area, was almost overwhelmed with opportunities to identify, categorise, sketch, and collect samples of plants, and to make ethnographic discoveries and notes as well.

The Pacific Northwest

One day, sailing in the Alaskan waters near Prince William Sound, Archibald made a surprising discovery: despite being the first Europeans to meet with the Indigenous people in this specific location, he was amazed to discover a man carrying

> [a] short warlike weapon of solid brass – somewhat in the shape of a New Zealand Pata-pattoo, about fifteen inches long; it had a short handle with a round knob at the end and the blade was of an oval form, thick in the middle but becoming gradually thinner towards the edges and embellished on one side with an Escutcheon inscribing Jos: Banks Esq.[85]

It was in fact one of the same brass pata ones that Banks had created and given to Cook for him to present as gifts in the region ten years earlier. Archibald was no doubt bemused to think a weapon from New Zealand, fashioned into brass in London, would now be found in a remote corner of the Pacific Northwest.

The discovery of Banks's pata may also have impressed upon Archibald how quickly the world was transforming into a global network of trade – not just between communities but now also between continents and cultures. The discovery motivated him to carefully observe everything he saw, just as his professor and mentor John Hope had advised.

Archibald's first encounter with Indigenous people of the Pacific Northwest was with the Nuuchahnulth people he met at Nootka in July of 1787. From there he went on to meet Indigenous people at Cheesish, Mooyah Bay, Nasparti Inlet [Port Brooks], Rose Harbour, and other areas nearby; he also sailed the Houston Stewart channel in the territory of the Kunghit Haida. With

Colnett, Johnstone, Taylor, and others he was among the first Europeans to initiate or continue first contact with Tsimshian, Heiltsuk, Haida, Tlingit, Eyak, and Yakutat peoples and their clans.

A journal entry by third mate Andrew Taylor describes the time Archibald tried to determine if the Indigenous people also considered Sundays to be a holy day:

> This being Sunday we got no fish, which was the Case last Sunday which led us to believe they paid some attention to the Day in order to discover whether this really was the case, the Surgeon with a Boy in our Canoe, went on shore to the Village. He was received very cordially by the Natives there was about two hundred present, they placed him on a clean Mat, by the side of Oughomeize the Chief of this district.[86] He remained with them an hour without discovering any religious ceremony or anything worthy of notice, unless 'twas one Woman endeavoring to relieve a sick Child of its pain by friction. this she performed with affection and tenderness, at the same time singing a doleful song.[87]

In early August of 1787, the *Prince of Wales* and *Princess Royal* met up with the previous two ships sent by the King George's Sound Company, and Archibald finally also met Captains Portlock and Dixon. The senior members of the four ships of the King George's Sound Company exchanged notes and updated each other on their fur trading exploits, their charts, and their relationships and knowledge regarding local tribes.

The geography of the rugged coastline likely reminded Archibald of his previous expeditions around the coastal islands of Scotland, and throughout this time he eagerly set out on the ships' small boats exploring places as far north as Lyell Island and visited Haida settlements at Port Yuka and At'ana. From there he sailed east to the south end of Banks Island, stopping at a small bay which Colnett named Port Ball and stayed there for 11 weeks. Here, the crew aboard the *Prince of Wales* made first contact with the Gitxaala Tsimshian, which was to be their most extended interaction with any of the people on the West Coast.

First impressions were made on both sides of these encounters. Writing to a friend in Edinburgh, the second mate on an unlicensed trade ship from Macao called the *Sea Otter* wrote a short sketch of the people he encountered around Nootka. This account was written in August of 1786, the same time that Archibald and the *Prince of Wales* were sailing out of London, but it would not have been any different had it been written 11 months later when Archibald arrived on the scene. The second mate, a Mr Elliot, gave his first impression of the Indigenous people:

> They are of a middle statue, stout and muscular built, of a dark copper colour, and very fierce countenance, which they heighten by smearing their whole body with red clay, mixed with fish oil, and streaking their faces with black; they also daub their hair with clay, and powder it with white down of birds. The dress of the men is a skin or matt (which they work of stripes of some tough bark very neat), tied round the neck; and open before, with a hat made of stained twigs, like basket-work, of the form of a sugar-loaf. The women wear a hat also, and the skins tied round their middle … but everything about them stinks most abominably, owing to the fish oil and other nastiness, with which they adorn themselves and their clothes.[88]

He went on to observe,

> Fish appear to be their chief provision in summer, and indeed they dry a great quantity
> for winter; they eat the whale, porpoise, and everything they can catch; I have seen them
> frequently take a mouthful of a fish even before taking it off the hook; in winter they have
> sea otters, seals, bears, wolves, and sundry other wild beasts, but no vegetable, nor any
> substitute for bread.

… and added rather remarkably, 'they likewise eat their captives and enemies killed in battle,
and I believe their own dead also; as, on our first arrival, they brought us for sale a great number
of skulls, hands, feet, and other human bones'.

Mr Elliot's modest ship (she was not much longer, at just 15 metres, than many canoes in the
area), the *Sea Otter*, was the first ship from Asia to visit the area.

When Archibald encountered local people for the first time, he would also have seen how
they painted their bodies. Maybe it reminded him that the earliest inhabitants of his own home
country were called Picts (painted people) by the visiting Romans because of the way they painted
and tattooed their bodies. The totems at the long houses may have reminded him of the standing
stones he grew up with, or of the heraldic crest carved in stone over the entrance to Castle
Menzies. It's possible his Highland background helped him to understand intuitively the nature
of the tribal social structure, the role of the chief, the way resources were distributed, and the
roles assigned within the community. He may have also understood that political allegiances and
tensions existed within the tribe, between tribes, and between cultures just as they had in his own
native land. His familiarity with the Highland ideals of *dùthchas* and *oighreachd* likely shaped his
perception of Indigenous culture in ways his English shipmates could not comprehend.

Another contact account, recorded by William Beresford, an assistant trader sailing on board
the *Queen Charlotte* with Captain Dixon, described an encounter as follows:

> In one of the canoes was an old man, who appeared to have some authority over the
> rest, though he had nothing to dispose of: he gave us to understand, that in another
> part of these islands, (pointing Eastward) he could procure plenty of furs for us, on
> which Captain Dixon gave him a light horseman's cap: this present added greatly to his
> consequence, and procured him the envy of his companions in the other canoes, who
> beheld the cap with longing eye and seemed to wish it in their possession.[89]

Beresford then went on to describe a trade negotiation with one of the women in the canoe:

> There were likewise a few women amongst them, who all seemed pretty well advanced
> in years; their under lips were distorted in the same manner as those of the women at
> Port Mulgrave, and Norfolk Sound, and the pieces of wood were particularly large. One
> of these lip-pieces appearing to be particularly ornamented, Captain Dixon wished to
> purchase it, and offered the old woman to whom it belonged a hatchet; but this she refused
> with contempt; … basons, and several other articles were afterwards shewn to her, and as
> constantly rejected. Our Captain began now to despair of making his wished-for purchase,

and had nearly given it up, when one of our people happening to shew the old lady a few buttons, which looked remarkably bright, she eagerly embraced the offer, and was now altogether as ready to part with her wooden ornament, as before desirous of keeping it.[90]

The Indigenous people also had their own unique take on the earliest encounters with these strange new people like Archibald with his blue eyes, ginger beard stubble, and shoulder-length light brown hair held together with a ribbon at the back. One such account comes from people living on the south end of Pitt Island at the time of first contact with the crew from the *Prince of Wales* and *Princess Royal.*

The story, passed down orally for 150 years, goes like this:

> One day, two Gitrhala men set out from their village to fish for halibut and were so absorbed in fishing that they failed to notice a large boat approaching. When one of them looked up, he saw a huge being with many wings approaching towards them. They at once thought it was a monster which lived in the nearby rocks. They were at the time fishing over a spenarnorh ('abode of a monster') from which a huge Raven used to emerge (a crest of 'Arhlawaels, Kanhade). They thought that the monster had now taken a new form and was approaching them to do them harm.[91]

This 'new form' was one of the ships' small boats – the very same kind that Archibald and Johnstone would use to row or sail from ship to shore to introduce themselves to local people. By the time the boat reached the shore, one of the Gitrhala men had hidden in the trees but one man remained, caught up in the ropes attached to his canoe. According to the story he passed on to his tribe later that day, the visitors

> … pointed first to the canoe and then to the halibut, and then to their mouths. This they kept on doing while saying 'Soap', which they gave him and took a halibut from the canoe. The Gitrhala man thought they were giving him a name. He was startled by the steel knives they used to cut the fish – so different from the sharpened Albatross bill he was using – and literally passed out when he saw them use the sparks from their flintlocks to start a fire of dried moss. He watched them boil water and add the fish and rice – which he had never seen before so identified as maggots – and then gawked in horror as they poured molasses on top – which he identified as human rot. He marvelled further as he watched them eat some biscuits … which he believed to be some form of tree fungus.

> Later, when the two Gitrhala men arrived at their home they told their people what they had seen and where they had left the strangers. Then the one Gitrhala man said 'They gave me the name Sabaen'. This is what it sounded like to him, but the white man was thought to have said Soap, because that is what he gave them. They kept it for some time before they found out what it was used for.

Were Archibald and Johnstone the two strangers in this tale? Possibly.

A recurring complaint arising from some of these early encounters, the source of much misunderstanding – and occasional violence – related to property ownership. The British traders were puzzled by instances of what they saw as theft. Sometimes, as when Colnett observed Chief Oughomeize intercepting canoes and taking a percentage of the catch, the perceived theft was between Indigenous people. Sometimes, as when Indigenous people were seen taking things from the ship or from seamen, the supposed theft was between the Indigenous people and the British.

In British eyes, such acts of 'theft' were deprecated as a sign of low morals. Had they understood the Indigenous coastal culture better, they would have known that chiefs had the right to everything within their territory – whether on land or even if it arrived by sea. Anyone entering their territory, European or Indigenous, risked seizure of their goods.[92] The Indigenous leaders were equally confused. Had they understood British culture better, they would have known that although there was a hierarchy on the ships, the hard-working, modestly dressed regular seamen were not slaves of the captain, and had property rights of their own.

These accounts illustrate some of the culture shock and confusion experienced on both sides of an early encounter. What they seemed to share was the opinion that strangers look very odd on first impression, utter sounds hardly anyone can understand, use bizarre technology, have odd beliefs about personal property, and probably eat their enemies.

Europeans saw canoes decorated with whale teeth and believed them to be human teeth. Indigenous people saw white-painted wooden rigging blocks and believed them to be human skulls.[93] Yet, despite the awkwardness, fears, and anxieties of these encounters, relations were generally good and over time improved when mutual trust was earned.

For the King George's Sound Company, trade was conducted peaceably, and commercial interests were advanced. Barkley had beaten them to Nootka, but they made good progress trading further to the north and in and around various coastal communities. Colnett began to believe the voyage might not be a commercial disaster after all.

For Archibald, expeditions to shores along the Pacific Northwest were the best days of his life. As surgeon he was constantly busy attending to the various ailments and injuries of the crew. Hernias were common; also broken bones, burns, cuts, and other consequences of life jumping in and out of boats and climbing over rocks; hunting and fishing accidents; gunpowder burns, and so on. As a botanist, Archibald's days were alive with the administration and discovery of knowledge. From the Argentine Isla de los Estados, where they had stopped for three weeks, he sent a tub of 20 small plants and ornamental trees to Banks and Hope. From every place he went, Archibald sent a steady stream of seeds, plants, and written notes describing the plants and their environment back to Banks by whatever means possible. He scientifically identified several plants, like the Nootka cypress tree (*cupressus nootkatensis*) and the Nootka rose (*rosa nutkana*) and over the course of the *Prince of Wales's* two seasons in the Pacific Northwest region (July–November, 1787 and April–August, 1788) he made great strides filling the gaps of knowledge in Banks's herbarium and gathering samples, specimens, and artefacts for Banks, for Aiton at Kew Gardens, for Hope at the Royal Edinburgh Botanic Garden, and for new friends at the British Museum. He even gathered an arbutus tree (*arbutus menziessi*) seed for Sir Robert to plant back at Castle Menzies.[94]

Over the three years of this mission, Archibald fulfilled his promise to Banks and sent back more than 100 specimens, including details on where they were gathered, and provided notes on each sample's scientific identity. Although half a world away from London, he continued to burnish his reputation in that city, and among its scientific community, as an able and accomplished natural scientist.

At the end of their first fur trading season (in November, when the snow began to fall), Colnett very sensibly stopped trading for furs and set sail for Hawaii.

Hawaii

Much had happened by the time Archibald reached Hawaii[95] for the first time. He had been away from London for 400 days, had survived scurvy, kept the crew alive, met many new and remarkable people, been exposed to new languages, met important tribal chiefs, and witnessed and explored the rugged coastline of the Pacific Northwest.

Now as the *Prince of Wales* and *Princess Royal* wintered at the mid-point between the Pacific Northwest coast and China, he would relive many of those experiences again.

However, the King George's Sound Company's mission here was substantially different. They were not in Hawaii to trade for furs but merely to rest and survive the winter – from January to March and again in September of 1788 – about four months in total. His first experience of Hawaii began on the eastern end of the group of islands and quickly worked its way west from Hawaii to Maui, then to Molokai, O'ahu (Honolulu), with most of their time being spent around Kauai and Puuai.

As in the preceding months, Archibald was employed equally as a surgeon and botanist. The geography and culture of the Pacific Northwest and Hawaii each provided Archibald with a host of new and exciting experiences. But the Hawaiian Islands differed significantly from the Pacific Northwest in many ways.

A description of Hawaii, recorded in a letter by Mr Elliot of the *Sea Otter* a year before Archibald arrived in the area, painted the scene:

> It is in length about ninety miles, and nearly as broad, of remarkably fertile soil, and very populous. It forms a delightful prospect, with its fine plantations rising gradually from the sea to pleasant green hills, and behind them mountains covered with snow … We found it exceedingly hot.

He went on to describe the people:

> The natives of this island were with us all day long, selling hogs, sweet potatoes, plantains, cocoa-nuts, bread-fruit, sugar-canes, salt, &c which greatly refreshed our crew. They are very tall, stout built, active looking people, many of them being seven feet high, and well made. Neither the men nor women wear any other cloathing than a maro, which is made of the rind of a tree, and narrow; this they put round their waist, and between their thighs, except on great occasions, when they wear a cloak … wrought with feathers, of a bright scarlet and yellow colour, as thick upon a bird. They also wear a helmet made exactly in the form of the old helmets of wicker work, and covered with feathers of the same colour as the cloak.[96]

Archibald was experiencing Hawaii for the first time but, as in the Nootka area, many of the Indigenous people he was meeting had received visitors before. British navy Captain Samuel

Wallis – the first European to land at Tahiti – had been here 20 years earlier, in 1767. French Captain Louis-Antoine de Bougainville – with whom Archibald had exchanged cannon fire at the Battle of the Saintes – had been here in 1768. Others, like Strange and Hanna had made pit stops here on their way to the fur trading missions in the Pacific Northwest. This was also the first visit to Hawaii for Colnett. He had served under Cook on the latter's second voyage and, keenly aware that Cook had met a violent death on the beach of Kealakekua Bay in 1779, made sure to avoid that part of the big island.

The dark cloud of Cook's fate aside, Hawaii was in many ways still regarded as something of a sunny paradise. For single young men – the average age of the crew being about twenty-five – who had spent weeks at sea, sometimes fearing for their lives, and who had been busy navigating the coast of the Pacific Northwest looking for opportunities to trade for furs, this three-month sojourn to the South Pacific seemed at times a heavenly reward for their efforts. The fine weather, the fresh fruits and produce grown in the lush fertile soil, restored their physical health without any prescription by Archibald. But the friendliness and beauty of the women sent their spirits soaring to new highs. Colnett was liberal-minded about shore leave and on-ship visits by local women. By 2 January, women were admitted onboard 'and every Sailor had a Lady in his burth'.[97]

Unfortunately these friendly relations may have meant that, while Archibald didn't have to treat anyone for scurvy, he had to be prepared to treat people for venereal disease. It is unlikely Archibald would have known that the introduction of venereal diseases to the islands could be traced back to the epic global voyage of Captain de Bougainville, the first Frenchman to lead a voyage around the world. Sailing with two ships of over 300 personnel[98], Bougainville called at the island in April 1768 after stopping in Rio de la Plata (presentday Uruguay).[99] As with scurvy, the effects were well known, but the cure was still something of a mystery. The same year that Archibald was in Hawaii pondering this disease, the distinguished Scottish surgeon John Hunter published his book *A Treatise on the Venereal Disease*, in London. But Archibald would not be able to read it for another year. For now, he did his best with the knowledge he had, treating the afflicted with doses of mercury, arsenic, and sulphur from his medicinal kit.

One cannot know for sure, but it is unlikely Archibald 'had a lady in his burth'. Apart from being responsible for the crew's health, he was (at age thirty-five) the oldest man on the voyage and also the most mature in attitude. He also had a healthy respect for women, and his relationships with them helped advance his scientific agenda and keep the peace.

When the *Prince of Wales* eventually made its way west from the main Island of Hawaii, past Maui and Molokai, the ship lingered ear O'ahu (Honolulu) Island for a week, and Archibald developed an important friendship with a woman called Nahoupaio, sister to Matua.[100] As the Nuu-chah-nulth women had kept an eye out for his safety, so too did Nahoupaio. His efforts to learn the local language, noted by Colnett in his journal,[101] not only helped his personal relationship with Nahoupaio but also helped reduce the risk of misunderstandings and conflict. One day, Nahoupaio told Archibald that some warriors were planning an attack and take over both the *Prince of Wales* and the *Princess Royal*: her early warning allowed the ships to avoid a conflict and undoubtedly saved many lives.

The sincerity of Archibald's efforts to learn the local languages, and the care and integrity with which he managed his personal interactions, endeared him to Kaeo, chief of the Kauai Island. The islands were not politically united at this time and Kaeo was one of several chiefs vying for dominance. Their friendship would prove to be an important one later, as Kaeo was the father of Kamehameha – the man who would unite the islands by conquest in 1795, and who would later to be known as King Kamehameha I. Thanks in part to Archibald's actions, Kaeo also became a great favourite both with the officers and crew. Archibald described him as moderate in stature, well-shaped, with mild regular features 'and a firm steady deportment well becoming his high rank as a king and great warrior'.[102] He was also easy and familiar in his manners, keen and quick in his comprehension and of a cool moderate temper. 'In matters of any importance, Kaeo was extremely inquisitive and deliberately weighed every circumstance so as not to suffer himself to be led astray by false appearances'.[103] He

> … would frequently seat himself near any of the mechanics that were at work, as he appeared extremely anxious of becoming acquainted with the principal modes of working iron and wood into various forms, so that if the blacksmith was doing anything in the forge, if the armourer was cleaning or taking the arms to pieces, or if the carpenters were doing any little job, he was their constant attendant, and seldom left the spot till they had finished whatever they were about.[104]

Another memorable person Archibald met, with Taylor, was a man named Tapowynah. He had apparently been a witness to Cook's death and had somehow obtained his blood-stained death shirt. Taylor described the encounter in his journal:

> This Morning Tapowynah brought off with him a Shirt of Capt. Cook's which he inform'd us Capt. Cook was Kill'd in. The Shirt had been Carefully taken care of since that Period there was Blood on it in several Places. … on one shoulder of ye Shirt he shew us is the place where He was stab'd with a Knife. & shew ye manner ye Natives Closed on him … In the lower part of yee Shirt on ye Fore part was another Slit which He said was occasioned by ye first Stab on his Groin or lower part of his Belly.[105]

This was perhaps Archibald's first experience as a forensic doctor, and it would have been rather chilling to hold in his own hands the death shirt of one who had inspired him to a life of adventure and discovery. He was no stranger to blood, but this artefact was also a stark reminder of the risks of the occupation and lifestyle he had chosen to pursue.

The dangers in Hawaii were as real as the pleasantries, and together they provided Archibald a roller coaster of experiences. One day he saw a sailor from *Princess Royal* taken hostage as a bargaining chip to be exchanged for arms and gun powder … the next, he heard the sailor had decided to stay behind permanently to live with the chief's daughter. Hostage one day, son-in-law the next. On another occasion in Hawaii, Archibald saw Colnett's ships repelled by armed warriors in nearly 1,000 canoes. Yet, another time they were greeted by canoes full of young women keen to come aboard and rub noses. Still another time – one of the more amusing episodes – Archibald and a few other members of the crew were asked politely by a curious chief to take their clothes off so their white skin could be more fully inspected.[106] He merrily stripped down and revealed all. There were moments of fear, anxiety, violence, and regret; but there were also moments of joy, humour, innocence, tenderness, and friendship.

Archibald was undoubtedly disheartened by the physical violence he witnessed on several occasions and was called upon to treat battle wounds more than he had hoped would be necessary on a trade mission. But he accepted that he was the lone intellectual among men who lived a rough and uncertain life and did not alone have the power to manage the behaviour or attitudes of others. Fortunately, the human desire to trade, to prosper materially, and to gather new knowledge, was strong enough to see all parties past their troubles. Archibald may have considered his experiences proved Adam Smith's assertions that people everywhere work towards their best interests.

With these first encounters in trade and enquiry completed, the next step for the voyage was to undertake its final accounting. The *Prince of Wales* met up with *Princess Royal* in Maui and on 30 September 1788 they set sail for Macao, arriving there six weeks later with 15 cases and 19 casks of furs.[107]

China

Macao, a Portuguese colony on the Pearl River, was an important and thriving port for Pacific trade located 50 miles downstream from Canton. Crafty maritime traders – such as Captain Barkley – would sail under a Portuguese flag to avoid the monopolies of the South Sea Company and the East India Company; and since Canton had restrictions on residency for some Europeans, Macao was a popular place to conduct business.

In the end, the King George's Sound Company's commercial venture was not as profitable as anticipated – launching a new enterprise in a new market proved challenging. Nevertheless, it was profitable for the business partners. Portlock, arriving a year earlier, had sold at least 2,000 furs in Canton earning a total of 80,000 Spanish dollars (or, according to Etches, about 21,000 pounds, perhaps $US 2.75M today).[108] The furs from the *Prince of Wales* brought in another 64,000 Spanish dollars.

After so many days at sea and among small villages, Archibald would have been thrilled to explore the thriving ports of Macao and Canton. For the last 30 years (since 1757) China – open to trade but not very keen on foreigners – had had a policy focusing all trade with the West through Canton, but it also required all ships to stop first at Macao, a port the Portuguese had been renting, more or less, from China since 1557. Both cities were emerging and active centres of commerce, and Archibald must have been thrilled to see mainland China for the first time. On the passage from Macau, up the Pearl River to Canton, he may have gazed with the keen interest and analytical eye of a botanist on the densely populated agricultural lands and the market towns they passed along the way.

Once in Canton, Archibald would have had an opportunity to meet other British traders as well as those trading with the Dutch East India Company and the Swedish East India Company. Despite the restrictions on European residency, there was a bustling and established European community here. The Swedes had even established a Masonic Lodge at their factory in Canton in the 1760s.[109] British ships were unloading silver, wool, and cotton – or in Archibald's case, fur – and loading up with tons of tea, silk, porcelain, and furniture. Small boats from the region, from large junks to smaller sampans, mingled with these vessels from around the world.

It took two months to get all their business affairs settled in China, and Archibald was prepared for the final leg of a voyage around the world that would present him with even more opportunities to expand his botanical achievements. He would pick up plant specimens along the way home with stops in Sumatra, Singapore, Martinique, Madagascar, and the Cape of Good Hope. However, as with their arrival in Nootka months before, their departure from China would not go as originally planned. Instead of everyone on the *Prince of Wales* and *Princess Royal* sailing back to England together, just as Portlock and Dixon had done the year before, the two ships split up. Colnett transferred command of the *Prince of Wales* to Johnstone, so he could take Archibald and a portion of the trading profits back to England right away. Then Colnett stayed behind to take command of a small fleet of four ships – the *Princess Royal*, the recently purchased 120-ton *Argonaut*, the *Iphigenia*, and the schooner the *North West America* – and set off immediately back to Nootka for another season of fur collecting.

The reason for this decision was both strategic and commercial. The month after arriving in Macao they met another trader by the name of John Meares. Until now, Meares had been a competitor of theirs, and a not entirely scrupulous one either; nevertheless, he now became their ally. Meares, like Captains Hanna and Barkley, had been trading in the Nootka area the previous year under a Portuguese flag and without licence from either the South Sea Company or the East India Company. Meares and Hanna had both been part of a Chinese-based company led by John Henry Cox. Meares had just returned from Nootka on his trading ship *Felice* and reported that the Americans were now elbowing in on the scene. The Boston trading ships *Columbia* and *Washington* had arrived in Nootka in September. Everyone agreed the situation was becoming messy. The two pairs of King George's Sound Company ships, the two ships sent by Strange, the two ships sent by Cox, had been stepping on each other's toes, driving up fur prices in Nootka and lowering profits in China by flooding the fur market. And now it looked like there would be even more competition – from the Americans. So, they decided to form a new company and return to Nootka in force, and under overall command of one person: Colnett.

Colnett summarised the situation in his journal:

> It was thought advisable by both parties to form a Junction of trade under the British Flag, each flattering himself [that] from the knowledge acquired by their Commanders of the Coast, [of the] dispositions of the natives, and articles coveted by them in Trade, [we] would soon expel all other adventurers, and enable us to make returns adequate to [meet] expenses of outfit which none of our former Voyages had done.[110]

He would take with him a team of 29 Chinese carpenters, blacksmiths, bricklayers, and masons to help build a permanent trading base in Nootka.[111]

The merger between the two companies was signed and sealed on 23 January 1789, and they set sail for Nootka on the 26th. Archibald and Johnstone made final preparations aboard the *Prince of Wales* and left Canton for London a few days later, on 1 Feb 1789.

Five months later, Archibald and the *Prince of Wales* were days away from home. After a not-unchallenging round-the-world voyage lasting three years, only one crew member had been lost. Archibald noted the mission's only death occurred just 20 days before they arrived back in London – the result, he explained disapprovingly, of a lingering disease contracted in China 'in consequence of intemperance'.[112] As ship's surgeon, it was not only commendable that his crew made it home alive, save the one casualty of over-indulgence in Canton, it was also a personal achievement to have survived. Others before him had not been so lucky: the first trained surgeon to visit the Pacific Northwest, William Anderson, surgeon of HMS *Resolution*, on Captain Cook's third voyage, had died of tuberculosis while off the coast of Alaska.[113] Surgeon's mate John Mackay had been stricken with 'a purple fever' on James Strange's 1786 fur trading expedition and had been left behind – and barely survived. Several crew aboard Meares's ship, the *Nootka*, died of scurvy the same year. Meares's companion ship, the *Sea Otter*, was last seen in Prince William Sound in September 1786 – the entire ship and crew lost forever.[114]

Entering the English Channel, the end of his mission now literally in sight, Archibald may have taken a justifiable moment to reflect on his personal journey. He had begun the voyage, a promising and experienced naval surgeon and part-time botanist, but he was now ending it as an expert. He had met people from many different Indigenous tribes and communities between Nootka and Alaska, had met with Indigenous Hawaiians from several of their islands, and with Chinese traders. He had circumnavigated the globe, including stops in Indonesia and Africa.

The experience was all he had hoped it would be.

He could only wonder what opportunity for further adventure might possibly come next.

Six
Aftermath (1789–1791)

As the *Prince of Wales* made its final approach towards London on 14 July 1789, Archibald may have noticed a number of vessels making their way from the French coast with some apparent haste. When he at last set foot safely back on land, he would have learned that on this day, revolutionaries stormed the Bastille fortress and prison there. The mighty kingdom of France was falling apart.

Archibald knew, of course, that France had successfully harassed and attacked the Royal Navy during the American War of Independence, and that the war had been expensive. But he was probably surprised to learn how little France had been able to afford their war. The financial strain on the state had burdened its citizens with higher and ever-escalating taxes; then recently they had suffered a crop-devastating drought and a cattle disease. The citizens of France were demanding the rights and freedoms espoused by Americans – provoked and encouraged by two Americans who had drawn inspiration from Archibald's professors in Edinburgh: Benjamin Franklin, who had been the US Ambassador to France from 1776 to 1785 and Thomas Jefferson, who was still in Paris consorting with the Marquis de Lafayette when the Bastille was stormed. Familiar thought leaders like Thomas Paine also added fuel to the fire.

Archibald was likely amazed to learn and see just how quickly the victors had become the vanquished, how the poor were becoming powerful, and how quickly frustration and anger had led to violence. He may have found it deeply ironic that the French king, who had so vigorously supported the American revolution against the British king, was now on the receiving end of populist revolt. The next month he read with awe the newspaper account of at least 6,000–7,000 persons lying dead in the streets of Paris as the chaos of revolution spread through the court and city.[115]

But Archibald's priority upon his return was to reconnect with Banks. He set the news of current events aside and penned a quick letter telling of his discovery of the brass 'warlike weapon' with Banks's name on it and to let him know he had 'given your name to a cluster of islands round where we was then at anchor' to mark the event. He signed off promising that 'in the course of a few days I hope I shall have the honour of pointing out to you their situation and extent on a Chart I have made of the coast – as also presenting you with a few mementos'.[116] He kept his promise, and within days was able to reunite with Banks at his Soho headquarters and show him the chart of Banks Island, and to give a personal account of the voyage and its many aspects of discovery.

Archibald would have told Banks stories of the people he had met – the abandoned surgeon's mate John McKay, the husband-and-wife adventure team known as the Barkleys, the Hawaiian woman voyager Wynee, the dominant Nuu-Chah-Nulth Chief Maquinna – and of the customs he had observed, and of course the many plants that had been scientifically identified using the proper Linnaean taxonomy. The discussions and expositions likely went on for days and the two men were undoubtedly joined by various other experts and enthusiasts who daily visited at Banks's residence to ask questions, espouse theories, contribute ideas, or merely to learn or be entertained.

It was probably during one of these visits to Banks's home that Archibald met another fellow who would become a lifelong friend: James Edward Smith. At thirty years old he was just five years younger than Archibald but was already, like Banks himself, a magnet for information about anything, everything, and anyone interested in natural science. Smith had also been a student of John Hope in Edinburgh and had recently purchased the entire collection of books, manuscripts, and specimens from the estate of Carl Linnaeus for the princely sum of £ 1,000 (about $US 120,000 today). The two had a great deal in common and spent many hours together.

Archibald was no doubt thrilled to learn that Smith had just established a society dedicated to studying and disseminating information concerning natural history, evolution, and taxonomy – a hotspot for people like himself who were interested in cultivating the science of natural history in all its branches. Smith called it the Linnean Society. He introduced Archibald to the new society's first treasurer: Samuel Goodenough – pastor, tutor, and amateur botanist, and also leading entomologist to Thomas Marsham, the first Secretary of the Society and an expert on moths and beetles. He was also pleased to discover that Banks's librarian, Jonas Dryander, whom he had come to know from his frequent visits there, had also agreed to be the society's first librarian.

Archibald immediately liked Smith – he was both intelligent and convivial – and was impressed by his vast social network. While Archibald had spent his post-Edinburgh years sailing around the Caribbean, the Atlantic, and uncharted Pacific coastlines collecting plants, Smith had been just as active cruising the salons of Europe collecting scientists. He was regularly receiving correspondence at his 12, Great Marlborough Street address from leading doctors, chemists, physicists, botanists, and naturalists all over the world: from Pierre Marie August Brousset in Paris, Michel Espirt Giorna in Turin, Thomas Hope (third son of mentor John Hope) in Glasgow, Edmund Daval in Berne, and Johan Gustaf Acrel in Uppsala.[117]

Smith was excited to learn about Archibald's botanical ramblings and insights but was startled to learn about one of Archibald's passengers aboard the *Prince of Wales*: a sixteen-year-old Hawaiian lad named Tooworero.[118] Tooworero had joined Duncan on the *Princess Royal* when she stopped in Molokai in 1788, then had joined the *Prince of Wales* in China, electing to return with Archibald and Johnstone all the way to London. It was an incredibly bold act on his part, but perhaps no bolder than Wynee sailing away to Nootka with Barkley … or perhaps no bolder than many sixteen-year-olds itching to travel beyond the literal and figurative horizons of their youth. In any case, Archibald and Johnstone had taken a special liking to Tooworero and, discussing his fate on the journey home, had jointly agreed to take him under their wing.

Archibald told Smith that when they arrived home and sailed down the Thames, he and Johnstone had both pointed out to Tooworero the key landmarks along the shore and, when they reached the 42-hectare Royal Arsenal at Woolwich, they showed him the massive powder magazines and munitions foundry there. A little further on they finally stopped at Greenwich and brought Tooworero with them on a tour of a 120-gun battleship. The young man was amazed with the display of power, as any young man would be.

Archibald and Johnstone made sure Tooworero saw something of London before heading down to Plymouth with Johnstone, where the latter had been appointed superintendent of a division of ships. Archibald relayed the scene in a memoir:

> Tooworero spent his first winter and spring down at Plymouth under care and tuition of Mr. James Johnstone who commanded the *Prince of Wales* from China, and was soon after appointed to superintendent a division of ships in ordinary at that port. This gentleman's first object was to have him inoculated for the smallpox, which he underwent with little inconvenience, and then he was sent to a public school in the neighbourhood, where great pains were taken to learn him to read and write. The first it seems could not be accomplished, for though he soon acquired a thorough knowledge and pretty exact pronunciation of the simple letters of the alphabet, yet no power of art could carry him a step further and get him to join or mingle these different sounds together in the formation of a word. But in writing he made better progress, that is, he soon acquired a habit of copying whatever was placed before him with great exactness in the same manner he would do a drawing or a picture. Indeed, to the art of drawing in general he appeared most partial, and would no doubt in a short time make great proficiency with the aid of a little instruction, but in this uncultivated state of his mind, he seemed fondest of those rude pictures called caricatures, and frequently amused himself in taking off even his friends in imitation of these pieces.[119]

It was at this time, catching up with Banks, Smith, and a host of other natural scientists, that Archibald learned his mentor and friend Dr John Hope had passed away in November 1786, shortly after he set sail with Colnett. It was a heartbreaking blow, but not unexpected; for having lived to reach the age of sixty-one Hope had had a comparatively long and full life. Eager to sustain Hope's work in Edinburgh, Archibald wrote in September to Hope's successor, professor Daniel Rutherford, advising on the seeds and botanical material he had collected on his voyages.[120] Rutherford sent Archibald's brother Robert, also gardener at the Edinburgh botanical garden, down to London to spend a few weeks with him and learn first-hand of his various discoveries and observations.[121]

Archibald wrote to Rutherford again in October to say he was making sure his brother was soaking up all the available knowledge for the benefit of the Royal Botanic Garden in Edinburgh and had taken him to visit Kew Gardens and its Director, William Aiton. While in London, they also visited the Scottish doctor and botanist William Pitcairn, for whom Archibald had collected specimens in the Hebrides years before. Pitcairn's Islington garden was 'so abundantly stocked with the scarcest and most valuable plants as to be second only in size and importance to Dr Fothergill's garden at Upton'.[122]

The two brothers enjoyed the time together, and Archibald was able to learn from him all the latest developments in Edinburgh and the news from Castle Menzies and its gardens. Robert may have told him about the up-and-coming Scottish poet Robert Burns. Possibly he even sang a line or two from Burns's popular ballad *The Birks of Aberfeldy*, written just two years earlier. Burns had composed the work following a visit to a place that Archibald knew well, and from which he had drawn his own inspiration when a young gardener in the Highland countryside. Robert sang in the poetic Scots language:

> The hoary cliffs are crown'd wi' flowers
> White o'er the linns the burnie pours
> And, rising, weets wi' misty showers
> The birks of Aberfeldie.

A new mission

Though he had only been back in London for a few months, Archibald was pleased to discover he was already being sized-up by Banks for another mission back to the Pacific Northwest. This new mission would be led by thirty-three-year-old Captain Henry Roberts, a skilled cartographer and a veteran of Cook's second and third voyages of exploration. Roberts had selected George Vancouver (whom he knew when Vancouver was a midshipman on the Cook voyages), to be his first lieutenant; other personnel assignments and details were still being put together.

Had he been so inclined, Archibald could have taken a short riverboat trip down the Thames to visit the Cuckold Point shipyard of Randall and Brent where construction of his next ship was completed. John Randall had been building ships for the Royal Navy for over thirty-five years and his partnership with John Brent, a well-experienced shipbuilder and assistant surveyor to the East India Company, had ensured a strong and steady business. Walking around the Randall and Brent yard, Archibald would have been energised by the hubbub of activity surrounding the construction of ships – the carpenters, rope-makers, sail-makers, blacksmiths. Many would be merchant ships sailing all over the world in the pursuit of trade, some would be sold to the Admiralty to protect those merchant and other interests, or for missions of research and discovery. After its initial construction was completed, the as-yet un-named ship was launched into the Thames and sailed to Portsmouth for final fitting and preparation.

Banks, impressed by Archibald's botanical work on the King George's Sound Company mission, was by then convinced beyond any doubt that Archibald should be the botanist on this new voyage around the world; he used all of his considerable influence to make sure of it. He also told Archibald to follow the newly launched ship down to Portsmouth and start preparing it for the botanical requirements of the voyage. He was determined that the ship be modified to include a glazed garden hutch on the quarterdeck and wanted Archibald to make certain it was built to suit his needs.

The visit to Portsmouth was also an opportunity to reconnect with his old friend James Johnstone; undoubtedly, they discussed the possibility of Johnstone joining the new voyage under command of Captain Henry Roberts. It was also an opportunity to check on Tooworero's academic progress. In an update letter to Banks, Archibald reported on the preparations for the botanical aspects of the voyage and asked that Tooworero be assigned to the voyage as his servant 'if not, I must provide otherwise for him'.[123]

Plans for the new mission progressed well, and on New Year's Day 1790 the 340-ton ship was formally purchased by the Admiralty. In tribute to the legacy of Captain Cook's earlier scientific voyages, and hinting strongly at the anticipated outcomes and prestige of this new mission, the ship was named HMS *Discovery*.

Less than three weeks later, boosting his pre-voyage morale and confidence even higher, James Edward Smith invited Archibald to become a Fellow of his newly formed Linnaean Society.

After eight months back in England, Archibald was becoming anxious not just about his own upcoming mission but also to hear news about the recently merged Etches and Cox companies. The four ships under command of Colnett had left China more than a year earlier, and enough time had passed for them to complete one trading season in Nootka. They should have returned to China with thousands of furs, and reports of their success should be making their way back to England. What had happened? Were they successful? Had they disappeared without trace in Prince William Sound, like the *Sea Otter*?

Crisis at Nootka

In April, news finally arrived. But it was not good. It seems the Spanish – who had long maintained a claim to the entire west coast of North America – stirred and awakened by the increased trading activity in the Nootka area, had decided to do something about it. Just as the Etches-Cox team had resolved to intimidate their American competitors by a show of force, the Spanish had also decided to play the heavy.

A Spanish naval force, sent from their base in Mexico and led by Captain Esteban José Martinéz, arrived in Nootka on 5 May 5, 1789 and raised the Spanish flag. When the Colnett party arrived shortly after, Martinéz had Colnett and his entire crew arrested and forced them to sail under armed escort to the military base Spain had established in San Blas Mexico, 20 years earlier. Martinéz also confiscated the *Princess Royal* and the *North West America* and, adding further insult, renamed them the *Princesa Real* and the *Santa Gertrudis la Magna*. It did not go un-noticed that the two American ships at Nootka, *Columbia* and *Washington*, were both allowed to remain there and carry on as before. As they had done during the American War of Independence, Spain was happy to side with the Americans if it helped to confound the British.

Colnett was put in irons but Meares, who had stayed in China to manage the business there, was spared from arrest. When word reached him about what had happened, he left immediately to personally bring word of these events back to London.

Unfortunately for the Spanish, Meares was quite an accomplished storyteller, and he added several dramatic flourishes to his narrative of events. Upon arrival in London, he gave his account to John Etches who then wrote to his older elder brother Richard, who in turn wrote to Banks:

> Sir, I am favored by my Brother with an Account of the arrival of Captain Meares from Canton, who brings official accounts of the seizure of all our Ships, Craft and the whole of the establishments on the N.W. Coast by a Spanish Admiral what pretensions they can have to such an act I am yet a stranger – except it is that monstrous, and absurd claim which they set up the last century 'An exclusive right to the Navigation Territories and Commerce of that quarter of the Globe'.[124]

Richard went on to explain the amount of risk and effort expended by his company and the huge profits that were just about to be finally realised,

Till of late we had innumerable obstacles to encounter, the chief of which we had surmounted and our Establishments wore the prospect of rewarding us with ample fortunes, a union form'd by my Brother when at China, with Captain Meares and his party had done away all competitors, had much enlarged our Capital, by subscribing a large sum for the furnishing an extensive Equipment of every kind of stores, and a reinforcement of people which was sent out under the unfortunate Captain Colnett …

News of Spain's aggression spread through informal and formal circles like wildfire. With each retelling, the story of the financial damages caused to the Etches-Cox enterprise escalated, and, since the newly formed private company was an evolution of the government-sanctioned King George's Sound Company, the acts inflicted by Spain were now deemed to be acts of aggression against Britain itself. Young Prime Minister William Pitt, age thirty-one and known as 'William Pitt the Younger' not just because of his youth but to distinguish him from his father (also William Pitt and also a former Prime Minister), was quick to seize upon Meares's claims, to bolster his status as an indignant and righteous defender of British sovereignty including the right of her merchants to establish trading posts and supply centres on any coast or island unoccupied by any other European power, and to defend freedom of the seas.

Pitt characterised the arrest of Colnett and confiscation of ships as acts of violence – an assault on Britain – and secured permission from King George III to demand 'immediate and adequate satisfaction for the outrages committed' by Martinéz and, if necessary, to back up those demands with a squadron of battle ships.[125]

All conversation at Banks's home, at London coffee houses, and taverns in Portsmouth, now focussed on the Nootka Crisis. The Spanish position, that places on land and various bodies of water belonged to whoever 'discovered' them first, was hotly debated, and countered by such as James Douglas (Lord Morton) who had advocated for Indigenous sovereignty even before Cook set sail, and others still, who argued that sovereignty was achieved by occupation – the establishment of a trading post, fort, or the like.

In the pamphlets and news sheets of the day, Archibald would have read of Pitt's sabre rattling with increasing alarm. He would also have read, and discussed with others, Meares's explosive 'Memorial to Parliament' after it was published in May. As one of the few people in England who had ever been to Nootka, and who knew both Colnett and Meares, Archibald's opinion and reaction to these topical events would have been highly sought after. He was likely torn between a desire to root for Team Britain, and a compunction to correct Meares's exaggerated claims about the level of infrastructure and trade he had established at Nootka prior to the arrival of the Spanish ships.

If Martinéz had intended to give British fur traders a rap on the knuckles, he had far over-achieved his goal. Britain reacted as though slapped in the face. Thanks to Meares's theatrical exaggerations regarding the losses to trade, infrastructure, and pride, there was now talk of war. Diplomats from Britain and Spain were urgently engaged in serious talks, desperately trying to avoid outright war.

Around this time, Archibald would also have heard that, after they bade farewell at Nootka, the Barkleys had possibly rediscovered the Strait of Juan de Fuca – which the great navigator Captain Cook had previously missed and had declared to be, like the mysterious Gallegos that had confounded Colnett, a Spanish folly. The consequence of this apparent confirmation that the Strait of Juan de Fuca existed caused people, including Richard Etches, Banks, and Pitt to wonder if there were other things Cook might have missed ... like a Northwest Passage to the Atlantic.

The possibility that a northern passage to the Atlantic might yet exist brought more official attention and interest to the Pacific Northwest than ever before. What if the Spanish found it first? It was no longer just privately funded traders and naturalists who were interested in the area. A region which, a few years before, was mostly of interest to fur traders and botanists was now of supreme interest to imperial politicians and policy-makers.

Even more shocking news arrived at this time, electrifying society's coffee houses and the establishments where naval surgeons, naturalists, explorers, botanists, traders, and adventurers met to discuss the news of the day: Captain Bligh of the HMS *Bounty* had arrived back in London.

Bligh's was an epic story of betrayal and survival in the Pacific. It is so well known today, largely because the tale of mutiny on the high seas captured the attention of all who heard it at the time. Everyone, including Archibald was alarmed that the greatest threat to the *Bounty* had come not from pirates, storm, or scurvy, but from within: from its very own crew.

Archibald would have been most interested in the fate of the *Bounty's* surgeon, Thomas Denman Ledward. Ledward was not only a fellow navy surgeon, but also a member of the Banks brotherhood of botanists. Banks had recommended Ledward for the position on the *Bounty* after receiving a letter from Bligh complaining that the official surgeon appointed by the Surgeon's Company, Thomas Huggan, 'may be a very capable man, but his indolence and corpulency render him rather unfit for the voyage'.[126] Bligh was not a doctor, but his diagnosis was astute – Huggan survived only to Tahiti, attaining the unique distinction of being the first Englishman to be buried there.

At twenty-seven years of age Ledward was not as old or experienced as Archibald but, then again, the mission was fairly straightforward: all they had to do was transport about 1,000 breadfruit and other plants from Tahiti to the West Indies. It proved anything but easy. After the mutineers took control of the ship, Ledward chose to risk death with Bligh rather than stay with the mutineers. He survived unbelievable odds at sea, travelling 3,800 miles in an open boat with Bligh; but was tragically lost at sea on his return home, after leaving Indonesia for Cape Town.

Archibald was equally excited and then saddened to learn the fate of Kew Garden's gardener David Nelson. Nelson, also tapped by Banks for the voyage, and a veteran of Cook's third voyage (on *Endeavour*), had survived the epic voyage to safety with Bligh and Ledward but died of fever shortly after reaching land.[127]

It must have been daunting for Archibald to face the stark reality, as he prepared for his next mission to the Pacific, that some of the surgeons, gardeners, and botanists tapped by Banks had had great adventures and made great scientific and ethnographic discoveries ... while others never came back alive.

In Portsmouth, as at Banks's home, Archibald was bombarded with news flooding in from the Pacific. The incident at Nootka, now an official foreign policy crisis, was uppermost in his mind. With each passing day it became increasingly likely that Britain would have to quickly counter Spain's actions and re-establish their right to trade in the Pacific Northwest. Spain may have claimed the entire West Coast but, in reality, had never done much to assert those claims and seldom ventured further north than the Spanish settlement of San Francisco, established by Gaspar de Portolà, governor of the Spanish province of Las Californias, in 1770.

George Vancouver

The Nootka Crisis reached such a crescendo that the Admiralty decided to re-assess Captain Roberts's voyage. It would now be a voyage not just of discovery but also of diplomacy; a decision was made to transfer the mission's leadership from Roberts to his first lieutenant, George Vancouver.

The new command was a heavy burden on Vancouver, but one he was eager to accept, and he had all the right credentials. Although just thirty-three years old, Vancouver had spent most of the previous 20 years at sea, serving first as a midshipman on *Resolution* during Cook's second voyage, and on *Discovery* during Cook's third. Returning from the Pacific, Vancouver quickly gained more experience serving short, but full assignments on a number of vessels. For six months of 1782 he was aboard HMS *Martin*, patrolling the English Channel and the North Sea before moving on to the clear blue waters of the West Indies. Once there, he transferred to the warship HMS *Fame* as fourth lieutenant with the British West Indies fleet in the Caribbean, briefly under Admiral Rodney, then under Admiral Pigot. When the American War of Independence was formally concluded, the *Fame* returned home, arriving in June of 1783, and Vancouver received his full commission to First Lieutenant on 10 July 1783.

The ship now placed under his command, the *Discovery*, was about 30 metres long and could hold a crew of 100 men. She was armed with 10 four-pounder cannon and 10 smaller swivel cannon – enough to defend herself but nothing like the 74 guns that on the 50 metre-long *Fame*. In addition to guns, and perhaps more importantly, the new ship carried the latest in high-tech navigational equipment: an Arnold Chronometer.[128] The chronometer was critical for measuring longitude, and those developed by Arnold were state-of-the-art. The one on *Discovery* was marked No. 176 and was an improvement over Arnold's innovative marine chronometer No. 3 which had accompanied Cook on his second voyage under the supervision of two astronomers appointed by the Board of Longitude. This new technology would help Vancouver produce coastal charts that were more precise, and more accurate, than ever before. If the elusive Northwest Passage were discovered by Vancouver, there would be no doubt how to find it again.

The change in command was exciting for Vancouver but briefly put Archibald off-balance – he was not entirely certain Vancouver supported the botanical aspects of the voyage or if he would make any other changes in personnel. But a flurry of correspondence between Archibald and Banks; Banks to Lord Grenville (the Home Secretary, and cousin to Prime Minister Pitt); and Grenville to Vancouver, eventually sorted everything out. It was quickly resolved that botany would continue to be a key priority for the voyage, and Archibald would remain in charge of that aspect.

However, for Vancouver it would be an unusual arrangement. Archibald would report not to him, but to Banks, and through Banks to the king. Archibald was a qualified Royal Navy surgeon, but his pay for this mission was arranged directly by Banks and approved by Lord Grenville. Furthermore, it was explained that when the voyage was completed, Archibald's journal, would be delivered to Banks, not Vancouver. Later, just before they set sail, Banks would write to Archibald again to underscore the lines of accountability:

> All the seeds of plants and the living plants you shall collect in your voyage you are to consider as wholly and entirely the property of his Majesty, and you are not on any account whatever to part with any seeds, plants, cuttings, slips, or parts of plants for any purpose whatever but his Majesty's use.[129]

Vancouver was instructed to support Archibald, provide him a boat and working parties as and when requested, provide him with trade goods for bartering with Indigenous peoples, and ensure fresh water was provided for the plant specimens he would keep on the ship's custom-built garden hutch.

Essentially, Archibald would be a member of Vancouver's crew, but not under his command. In fact, as to botanical aspects of the voyage, Archibald was authorised to command Vancouver.

Fortunately, Vancouver – generally regarded as a rather inflexible and not all that interested in natural sciences – accepted the situation. He wrote in his journal,

> Mr. Archibald Menzies, a surgeon in the royal navy, who had before visited the Pacific Ocean in one of the vessels employed in the fur trade, was appointed for the specific purpose of making such researches, and had, doubtless, given proof of his abilities, to qualify him for the station it was intended he should fill. For the purpose of preserving amongst His Majesty's very valuable collection of exotics at Kew, a glazed frame was erected on the after part of the quarter-deck, for the reception of those he might have an opportunity of collecting.[130]

The 2-metre by 4-metre garden hutch was an unusual feature, and Vancouver seemed to understand how important it was to the mission when he assumed command of *Discovery*. However, he was more interested in two other aspects of the voyage that were his own personal responsibility: diplomacy and geography. Vancouver's principal tasks were to officially receive back, on behalf of Great Britain, the properties seized at Nootka by Spain; he was also to court and woo Hawaii's Indigenous leaders so that Great Britain could continue to use those islands as a strategically important supply base; and he was to explore and chart the west coast of North America from California to Cook Inlet, Alaska – particularly with a view to discovering (or disproving) a Northwest Passage to the Atlantic, and to definitively determine the most northerly extent of the Strait of Juan de Fuca.

Archibald was pleased that Vancouver had a very full plate, and confident the good captain would be happy to leave botany to the botanist.

Meantime, Archibald frantically oversaw the hutch's construction, making sure it was completed before they set sail, writing to Banks to keep him apprised of every detail.

The plan had always been to send two ships on the voyage but, following the outrageous and incredible story of the mutiny on the *Bounty*, this was more important than ever. In this case the 135-ton ship HMS *Chatham* was assigned to accompany the Vancouver expedition. Twenty-eight-year-old William Broughton was appointed as her commander. Though young, Broughton had 16 years' experience in the navy, having sailed previously in North America, the East Indies, the Mediterranean, and most recently aboard the massive 104-gun HMS *Victory*. *Chatham* had been built a couple years earlier than *Discovery*, in 1788, but was still relatively new when acquired, and armed with four three-pounder guns and six swivels.

Smaller than *Discovery*, but with the advantage of a copper-sheathed hull, *Chatham* carried a crew of 55 men.[131] Archibald was pleased to learn that his friend Johnstone had managed to get himself appointed her master. Johnstone had also recruited fourteen-year-old Adam Brown to join him as master's mate. In the years ahead the three would spend many hours together on small-boat sorties and become lifelong friends.

HMS *Daedalus* was also assigned to the mission as a supply ship, to shuttle supplies, correspondence, and materials between the Pacific and England. Much of this would include correspondence and packages from Archibald to Banks, and which Banks would then redistribute to Kew Gardens, the Royal Botanic Garden in Edinburgh, and the British Museum.

In these final weeks before departure, the *Discovery*'s hold was loaded with stores sufficient to last a year and a half, and with beads, copper, and other items that could be presented as gifts – as a mark of respect or to curry favour – or to exchange for fresh produce and other supplies that might be required in any number of different circumstances. As Archibald has advised, they brought fewer beads to trade with and more practical items instead – copper, metal, axes, cloth.

The last eight months in England now passed by relatively quickly.

In July Archibald was briefly distracted from his preparations by the news that Adam Smith had died in Edinburgh. While a student there Archibald had been impressed by Smith's ideas and had given them considerable thought in the years since joining the navy and sailing around the world. Having since been to the Caribbean and other places, and having seen how slavery was used on plantations there and in the American east coast, and in some native communities, he had gained deeper appreciation for Smith's argument that free individuals, because they are working for their own best interests, work harder than slaves and are motivated to invest in the improvement of their resources in order to earn a higher income. He agreed that the 'invisible hand' of the marketplace was more effective at creating prosperity; that society was better and more productive when people were free to act in their own interests, rather than being coerced into action by others. He was grateful to have heard Smith's ideas circulating around Edinburgh when he was a student there and hoped in his small way to advance his ideas, so far as he could as surgeon and botanist.

More positive news arrived in October: a formal convention settling the issues of Pacific navigation and settlement was signed between the governments of Great Britain and Spain, giving Vancouver the green light to formally re-possess Nootka. Spain had been watching their long-time ally France descend into economic and social chaos and was no longer confident they could rely on her for support in armed conflict against Great Britain. Giving up claims on Nootka was far cheaper than war.

The best news of all arrived in March, when Archibald received a five-page letter of instructions from Banks confirming his assignment on *Discovery*. The letter was precise, complete, exhaustive, and definitive. There was no doubt now about what was expected from him.

Banks instructed that Archibald was to investigate the entire natural history of the countries visited, paying attention to the nature of the soil, and determine whether grains, fruits, and vegetables cultivated in Europe were likely to thrive there. All the trees, shrubs, plants, grasses, ferns, and mosses were to be documented and listed by their scientific names as well as the names used for them in the language of the Indigenous people. He was also to dry specimens of all that were worthy of being brought home and all that could be procured, either living plants or seeds, so that their names and qualities could be determined at the king's gardens at Kew. If he found any curious or valuable plants that could not be propagated from seed, he was to dig them up and plant them in the glass frame on the deck of *Discovery*.

As if that were not enough, his scope of work required that he examine beds of brooks, sides of cliffs, and search for ores or metals and minerals. He was also to note beasts, birds, and fish likely to prove useful either for food or in commerce. He was directed to pay particular attention to the natural history of the sea otter and obtain information concerning the wild sheep, and to note all the places where whales or seals were found in abundance.

Finally, he was to enquire into the manners, customs, language, and religion of the Indigenous people and gather information about how they manufacture items, particularly the art of dyeing. He was to keep a regular journal of everything noted and that journal ' … together with a complete collection of specimens of the animals, vegetables, and minerals obtained, as well as articles of the cloths, arms, implements, and manufactures of the Indians, were to be delivered to H.M. Secretary of State or to such person as he shall appoint to receive them'. [132]

That last point seems simple enough, but it would later be the source of considerable drama.

Lord Grenville sent a copy of these instructions to the Lords Commissioners of the Admiralty asking that they impress upon Vancouver that he was to afford every degree of assistance to Archibald in pursuit of these mission requirements.

Expectations and responsibilities were high, and Archibald was likely pleased and relieved to be assigned a botanist's lieutenant: John Ewin. The Admiralty also approved his request to include Tooworero on the voyage.

He was also pleased he'd have no duties as ship's surgeon, since A. P. Cranstoun had been assigned to *Discovery* as chief surgeon, supported by surgeon's mates G. C. Hewitt, Adam Mill, and J. Mears. The *Chatham* was to be supported by doctors William Walker and William Nicholl. With five physicians assigned to the two ships, Archibald was free to focus entirely on the botanical aspects of the voyage.

For an adventure-loving surgeon with a passion for botany, everything had come together wonderfully. Archibald was excited to set sail from Portsmouth again. Other exciting events were happening at this time too – Joseph Haydn was preparing to give his first performance in London, and Thomas Paine was about to publish his ground-breaking *The Rights of Man* – but such luxurious terrestrial experiences would have to wait till another day.

For now, it was time to pack up his bone saw and botany books and head back to sea.

Seven

NEW HEIGHTS IN HAWAII (WINTERS 1791–1794)

The night before leaving England for his second round-the-world voyage, Archibald wrote a short, perhaps final, letter to his mother, concluding with the heartfelt but slightly ominous words ' … may the guiding hand of Divine Providence long continue its protection towards you & grant you the full enjoyment of Health in this life, and happiness in that which is to come hereafter' signing off, perhaps forever, 'your dutiful son – adieu!'.[133]

The next day, still clearly affected by the ambitious scope of the mission and the knowledge he would be away for several years, Archibald reflected in his journal that he was leaving ' … not without those tender emotions which naturally occupy the busy mind on such occasions – in this parting for a time with our nearest and dearest connections – I am thus leaving our country at a moment when towering aspect of public affairs throughout Europe seems to indicate a general War'.[134]

What important events might transpire in the world of public affairs, and what fate might befall his family in Scotland, were as much a preoccupation as what might come of this voyage in the days, months, and years ahead. But Archibald was correct to sense that things at home and aboard the mission's ships would be eventful and have far-reaching consequences. To begin with, Lord Grenville, a vigorous champion of their mission, was elevated to the role of Foreign Secretary the very next month. This not only gave the voyage a stronger ally in government circles but also heightened the diplomatic importance of the mission. Failure by Vancouver and his crew to successfully de-escalate the Nootka Crisis, or to build strong and friendly relations with the Indigenous peoples in Hawaii and along the coastline of the Pacific Northwest, could have drastic international consequences. Equally important to British state interests was the task of finding – or definitively disproving – the existence of the Northwest Passage from the Pacific to the Atlantic.

The two ships, *Discovery* and *Chatham*, would sail around the globe in the opposite direction from that which Archibald had travelled before with Colnett on the *Prince of Wales*. They would sail first to the southern tip of Africa and from there, east to Hawaii. The west coast of continental America, they would explore in the summer months, with Hawaii serving as a winter destination.

On their mission, Archibald and all the senior members of the two ships' crews were mindful of Lord Morton's directive to Captain Cook many years before, which held that all the Indigenous peoples they would encounter were, like everyone else, creatures of God and are ' … the natural, and in the strictest sense of the word, the legal possessors of the several Regions they inhabit. No European Nation has a right to occupy any part of their country, or settle among them without their voluntary consent'.[135] Lord Grenville had also written to the Admiralty, before the ship sailed, advising that

> … in the execution of every part of their service it is very material that the commanding officer should be instructed to use every possible care to avoid disputes with the Natives of any of the parts where he may touch, and that he should be Particularly attentive to endeavor, by Presents and by all other means, to conciliate their friendship and confidence.[136]

Diplomacy and tact were critically important to the success of the mission, and while Vancouver would be ultimately responsible, the whole crew had a role to play in its success.

Archibald must have believed he could be a strong contributor to this aspect of the mission. He already had experience communicating with Indigenous peoples in the Pacific North and Pacific Northwest regions; he had established friendships with some tribal chiefs; he had Tooworero to help him in Hawaii; and he had spent countless hours observing, discovering, and documenting the natural history and local customs of the areas they were to visit.

Even so, there were still some mysteries and dangers to face. Just before leaving England Archibald received a private letter from Banks, written at the suggestion of Lord Grenville, on the matter of Maori cannibalism. There were credible stories of the practice dating back to a violent 1772 conflict between French explorers and Maori warriors, and Banks himself had seen mummified heads on his visit with Cook. But the existence or purpose of such a practice was still unconfirmed and had not yet been properly assessed by a trained naturalist. So, Banks advised that

> … if you find the abominable custom of eating human flesh, which they are said to practice, to be really used among them, you are, if you can do it with safety and propriety, to be present at some of their horrid repasts in order to bear witness … and discover, if you can, the original motives of a custom for which it seems impossible to suggest any probable cause.[137]

The diplomatic challenges ahead were both intimidating and exhilarating. Failure to establish good relations – caused even by a good-faith misunderstanding – could result in death, war, and loss of standing with these strategically important developing nations. It was a great deal of responsibility for what was, by present-day standards, such a young crew.

At thirty-six years of age, Archibald was one of a handful of senior crew in his thirties and the only person on the voyage to have spent any time at all in university. Virtually all the others on the mission, even the most senior, had spent their teenage years and their twenties at sea. George Vancouver, age thirty-five, had been in the navy since he was thirteen. Lieutenant Joseph Whidbey, also thirty-five, had been at sea since his teens. Twenty-nine-year-old William Broughton, in charge of *Chatham*, had been in the navy since age twelve. Lieutenants Zachary Mudge, Joseph Baker, and Peter Puget were all in their early twenties and had been in the navy since being ten or twelve years old. In fact, Baker had been a cabin boy at the Battle of the Saintes, and Archibald remembered him from days on *Formidable*. Baker had also previously served on HMS *Europa* where he had met then-Lieutenant George Vancouver and then-Midshipman Peter Puget, as well as Joseph Whidbey.

Archibald's friend Lieutenant Johnstone, now age thirty-one, had held out until he was twenty before joining the navy and his comparatively wide breadth of life experience was likely one of the things that Archibald appreciated about him – they were both committed to their Royal Navy careers, both seasoned in battle, but they had experiences outside and beyond the navy which added depth to their conversations and growing friendship.

The average age of the 145 persons aboard *Discovery* and *Chatham* was twenty-three; the youngest was sixteen years old and the oldest – the quartermaster aboard *Chatham*, Thomas Miller – was forty-five.

Generally speaking, the crew fell into four social classes: at the top were the officers, identifiable by their dress in white breeches, blue waistcoat and matching blue overcoats with brass buttons, and tricorn hats; then the warrant officers, not in uniform but dressed as gentlemen of the day; the next rung down were the teenage midshipmen, basically officers-in-training; and at the lower end were the hard-working able seamen, who dressed in their own choice of work clothes and performed the rough manual, though not unskilled, labour of sailing.

Archibald was accounted among the officers' social class, but in his role as botanist had the same liberty as when he had served as surgeon: he dressed in the style of a gentleman – which is to say he appeared in the same style and quality of dress as the officers but with more variety of colour. Brown breeches, white shirt, red waistcoat with a brown over coat was a popular scheme at the time; it's possible Archibald felt those earthy tones suited his line of work. Like the officers, he wore a tricorn hat. A true naturalist, he likely preferred to keep his naturally light-brown hair tied back in a ponytail. In keeping with his status, Archibald was assigned his own private quarters and dined with the officers in the officers' mess. Lieutenant Ewing acted as his assistant, and Tooworero played a similar support role.

In the warrant officer class were the ship's surgeon, A. P. Cranstoun, and surgeon's mates G.C. Hewitt, Adam Mill, and J. Mears. On *Chatham* the warrant officer class included Dr William Walker and his mate William Nicholl. They would all dine in the gunnery mess with others in their class such as the chief mate. Like the officers they acted as mentors to the younger midshipmen on board and, in addition to their own specific roles, they supported the officers by training the midshipmen for their admiralty exams. Midshipmen received lessons in seamanship: how to make bowlines, braces, backstays, bobstays, shrouds, buntlines, clewlines, tacks, sheets, earrings, reef tackles; and take lessons in navigation involving logarithms, sines, tangents, cosines, secants, cosecants, horizontal parallax, semi-diameter, meridian altitude, the heavenly bodies, planets, and stars.

The able seamen, in addition to being in excellent physical condition, were technically proficient tacticians and craftsmen skilled in sail-making, carpentry, brewing, cooking, blacksmithing, rope-making, and a dozen other skills necessary for survival and success at sea. The *Discovery* crew included sixteen marines; *Chatham* had eight.[138]

Pitt and punishment

Looking over the crew list Archibald may have noted that among the *Discovery*'s midshipmen were two particularly highfalutin sixteen-year-olds. One was Charles Stuart, the son of the Earl of Bute; but the other and far more troublesome one was Thomas Pitt. His father was the 1st Baron Camelford. William Pitt, the former Prime Minister, was a relative. Lord Grenville was a brother-in-law. These powerful family connections helped the adventurous young Pitt get his assignment on the *Discovery* voyage – a mission almost guaranteed to be of great historic significance. To be fair, he did have some previous experience at sea: as a fourteen-year-old member of HMS *Guardian*, under command of Captain Edward Riou – who had been a midshipman under Captain Cook on the original *Discovery* – Pitt had survived a near-fatal shipwreck with an iceberg.

The experience on HMS *Guardian* was the sort of near-death survival story that might have shocked any normally wild and rambunctious boy out of his fantasies of fame and adventure, but in Pitt's case seems only to have encouraged him further.

Initially Pitt got along well with Captain Vancouver and the other officers. He received daily instruction from lieutenants Baker, Mudge, and Puget. Archibald explained the basics of botany. Occasionally the captain himself would give a lecture on the finer aspects of navigation, of which he was one of the most accomplished in the world. Vancouver even promoted Pitt to master's mate two months into the voyage. But the relationship soured drastically when the expedition reached Tahiti ten months later. Vancouver would not allow anyone other than himself or his officers to trade or mix with the Indigenous peoples. He laid down strict rules about this and warned that he was not to be crossed. Aware of the dangers to the crew, and to the locals, caused by innocent mistakes or various misunderstandings, and mindful of Lord Grenville's directives, he also forbade any midshipmen or able seamen from going ashore unless under orders to do so. This was a huge disappointment to many of the young men, bedazzled by the nubile Tahitian women who paddled out to the ships enticing the men to trade.

The policy irked Pitt considerably and, convinced of his own superiority, he figured he could bend the rules. So, one day, when both ships were anchored at Matavai Bay on the north side of Tahiti, Pitt tested Vancouver's authority by tossing a piece of iron – a highly valued trading commodity – to one of the Tahitian beauties who had pulled up next to the ship. It was a simple act, perhaps even a spontaneous, playful, and innocent act. But there was no room on Vancouver's *Discovery* for insubordination. His rules and his authority, for everyone's safety and the success of the mission, had to be inviolable. Vancouver, like Archibald and everyone else, had read of the incident with the *Bounty* – men turned to mutiny, bedazzled by the prospects of life in paradise – and was determined not to see those events repeated under his command.

When word of Pitt's act reached Vancouver, he ordered Pitt to present himself in front of the officers and all the other midshipmen. Archibald could likely sense the awkward tension in the room as Vancouver demanded Pitt account for himself and may have been alarmed when Pitt stood firm in his denial of having done anything wrong. Lieutenant Mudge attempted to de-escalate the standoff but neither Pitt nor Vancouver would back down. Pitt, adamant and stubborn, would neither apologise nor admit wrongdoing. Vancouver would not show mercy. A punishment of flogging was ordered, and Pitt was strapped across a cannon and received, in front of all officers and his peers, two dozen lashes from the boatswain's mate. The ship's surgeon supervised. It was a humiliating lesson Pitt would not forget. Even so, over the next two years he'd be flogged twice more and demoted back to midshipman. Eventually, when he could tolerate him no more, Vancouver had Pitt transferred to the supply ship *Daedalus* for return to England.[139] The relationship which started so well was hell for them both.

Archibald had of course seen men flogged before, but he undoubtedly thought the dramatic standoff with Pitt in Tahiti provided a revealing insight on Vancouver's character and leadership. Before joining *Discovery* Archibald had likely known Vancouver mostly by reputation alone, but by the time they reached Hawaii he was beginning to see that Vancouver could be both cruel and kind in equal measure.

The rule about shore leave and indiscriminate trading would be upheld rigorously over the remaining years of the voyage and would extend to local traders as well. Vancouver would allow no visitors on board unless invited – an honour reserved mostly for local chiefs and kings – and he did not like crowds of people or boats to gather around the ships. In time people would see the wisdom of this policy, and Archibald would observe:

> While the natives are not suffered to crowd the vessels, the inducements to theft and misunderstanding are less frequent, and they felt less compunction in the restraint, and seemed equally happy and contented in being suffered to come alongside in their canoes and dispose of their articles peaceably.[140]

Spontaneous crowding and gatherings were a regular feature of the ships' arrival near shore and a source of both delight and concern. When anchored at Kealakekua Archibald wrote that the *Discovery* and *Chatham* were

> … daily surrounded by a great concourse of the natives in canoes, and a great many of both sexes who had not the means of coming off in any other way, came swimming alongside and continued gambolling in the water most of the day, but as none but the king and a few of the principal chiefs were suffered on board in the day time, the duty was carried on with much facility and no interruption had hitherto taken place of the most perfect harmony and good understanding which the king on all occasions showed a ready disposition to cultivate and preserve.[141]

The Hawaiian King Kamehameha also ordered some of his own canoes to paddle around the vessels at night to guard against any breach of protocol or good faith. Vancouver, bossy and demanding, had no argument about his policy from Kamehameha, who was just as happy to see the midshipmen and able seamen kept at arm's length from the general population – in everyone's best interests.

This hard-nosed approach not only preserved the peace – the 'perfect harmony and peaceful understanding' that Archibald observed – but also helped advance Vancouver's diplomatic goals in Hawaii. The approach also bore fruit both figuratively and literally. Archibald observed in his journal that

> Hogs and every kind of vegetable the island afforded were abundantly supplied as they were wanted by mentioning to Kamehameha the evening before what quantity should be brought, for he had taken in a great measure upon himself supplying both vessels, declaring that as they belonged to King George they must not in his dominions traffic for refreshments like other vessels, but be supplied during their stay with whatever they stood in need of in the most liberal way.[142]

Archibald had learned quickly that Vancouver, when offended – personally, or on the behalf of others, or on a matter of principle – was zealous in the administration of justice.

Another example of this trait occurred after Vancouver invited King Kahekili and King Kaeo to spend the night on *Discovery*, together with some other chiefs and their attendants. The next morning one of the women in the kings' party reported she was missing a piece of ribbon and complained to Vancouver that someone had stolen it. Archibald observed the scene as Vancouver grew increasingly passionate in his determination to have the ribbon retrieved, writing that he

> … threatened the chiefs with such menacing threats that he terrified some of them out of the ship with great precipitation. The king in particular came running into my cabin before I knew anything of the business and instantly jumping into his canoe through the port hole, paddled hastily to shore … [143]

While in Hawaii Archibald also witnessed first-hand Vancouver's unyielding desire to see justice meted out to those who threatened the ships or their mission. This was revealed to him when he, and Vancouver, learned that the captain of the mission's supply ship *Daedalus*, Richard Hergest, and her passenger William Gooch had been killed at Waimea Bay.[144] The news was particularly painful to Vancouver as he knew Hergest personally from when they had both served as midshipmen with Cook years before. He was also angry to lose Gooch, for he was supposed to join *Discovery* as her astronomer.

Losing Hergest and Gooch was a matter Vancouver could not let go unpunished and he eventually raised the matter with King Kahekili. Making note of the discussion between the two, Archibald later reported in his journal:

> Kahekili protested his innocence of the whole transaction, and expressed his indignation at the cruel circumstance of the murder, which he said had been committed by a fierce and ungovernable mob of common people who had no chiefs among them at the time to check or restrain their barbarous conduct. He readily allowed that neither Mr. Hergest not any of his party gave cause of offence by any kind of bad treatment, for which he had already ordered three principal murderers to be put to death, which he hoped would sufficiently prove his detestation of their conduct.[145]

But Vancouver was not satisfied with this and insisted there must be more than three men guilty of the murders. After some questioning of a witness to the event, and with damning testimony from Chiefs Kamohomoho and Kanawai, three men were eventually found, tried, and convicted. Archibald recorded the gruesome execution of the death sentence:

> The three criminals were then conducted on the quarter deck where the marines were drawn up under arms, and there put into irons, while the ship's crew were called up and made acquainted by Capt. Vancouver with the crime for which these men were now to suffer. After this our men were ordered to their quarters, and the two chiefs ordered a large double canoe alongside of the gangway into which one of the prisoners was carried and lashed down on his back, when Kanawi, who took upon himself their execution, went with a ship's pistol into the canoe and shot him through the head, so that he died without a groan or a struggle …

And the other two men met the same fate.

> When the disagreeable business was thus ended, the two chiefs came up to Captain Vancouver on the quarter deck and taking him by the hand, asked him if he was satisfied with what they had done … he told them he was perfectly satisfied with their conduct and assured them that all animosity on account of this business should now end.[146]

Learning from past transgressions and taking every effort to avoid new ones was a constant struggle for everyone. But the British officers and Hawaiian chiefs were able to participate in at least one joint act of reconciliation. At Kealakekua Bay, Vancouver, Archibald and some others from *Discovery*, accompanied by King Kamehameha and his queen, paid a visit to the place where Cook had been killed. Vancouver already knew the location well, for he had experienced his own 'severe mauling' not much farther away – at Napo'opo'o Beach, on the southern end of Kealakekua Bay – when the infamous event occurred. Though Archibald was seeing the site for the first time he would have remembered the story of Cook's death shirt, related by Andrew Tayor years before, and in his mind's eye could recall where the fabric had been punctured and torn by the knife's fatal blade.

> On landing, the chief of the district, an elderly man named Keawe-a-heulu, came down to the beach to receive us, and with a dejected air of condolence, he pointed out to us the fatal spot on which Captain Cook lost his life, and described with minuteness every circumstance attending that unfortunate and ever to be lamented event, in a manner that did great credit to his feelings, and pervaded the whole circle present with a gloomy aspect of mournful condolence. No people could show more regret in bewailing the death of our illustrious navigator than the natives did on this occasion, when the rash and precipitant circumstances of the whole transaction was thus brought fresh to their memory.[147]

Not only did Keawe-a-heulu show them the site of Cook's death, he also took a bold further step of trust. Archibald continued,

> To show the confidence they placed in the reconciliation and promise of the successor of Captain Cook that all revenge and animosity on account of this horid event should cease and be buried in oblivion, the very man who from cruel rashness stabbed him was now present, and pointed out to us among the multitude.

The party then took a short walk to stone wall, a shrine of sorts, and observed a number of wooden carvings of local atuas and gods before leaving. It was a poignant moment, and a welcome step forward in the relationship they were trying to forge.

Tooworero

In total Archibald spent about five months in Hawaii, visiting three times between the *Discovery's* first arrival in March of 1792 and its final visit in January 1794, and he enjoyed renewing and making new friendships during his botanical and other exploits there. But no one aboard

Discovery was more excited to see Hawaii again than Tooworero. It was four years since he first boarded *Princess Royal* on the King George's Sound Company mission and since then he'd seen the cities of Canton and London, toured battleships at Greenwich, been on a trans-Atlantic voyage to Hudson Bay, seen the world's largest naval base in Portsmouth, seen Cape Town, and walked the shores of Australia and New Zealand. By March 1792 he was happy to return home and it soon seemed any hardships he may have endured in the previous months were well worth it.

King Kamehameha, an emerging anglophile, recognised the potential for Tooworero to assist in relationship building with current and future representatives of King George, and elevated his social stature by rewarding him with a small plantation.

In the following months Tooworero would also marry a chief's daughter and, consequently, acquire a second plantation. Archibald reckoned he had 'at least two hundred vassals who considered him as their chief'.[148] Although he had not become proficient in reading and writing English, Tooworero's verbal skills were excellent, and the time he had spent with the Europeans at sea and in Portsmouth gave him valuable cultural context for many situations. Thanks to Archibald's mentoring and Johnstone's guidance, he was now able to act as a translator and interpreter, smoothing the way for the Europeans and the native Hawaiians to get along and avoid the sorts of misunderstandings that had cost Cook and others their lives. On return visits to the area during the Voyage of Discovery, Vancouver, Archibald, Johnstone, and others would welcome Tooworero and his wife on board the *Discovery* as frequent guests.

There was, however, a moment when Tooworero's rapid social ascent was very nearly short-lived. When Tooworero presented himself to Vancouver for the parting gifts he had been promised, Vancouver gave him a half dozen axes and some other items of lesser value. Tooworero was less than fully satisfied with these and wasn't the only one to think Vancouver was sometimes less generous than he ought to be. So he paddled over to try his luck with the more convivial Lieutenant Puget on *Chatham*, who duly responded by presenting him with two canoes and, as Archibald wrote later, 'other articles of utility and ornament as he thought would contribute to his comfort and happiness and tend to give him consequence among the other chiefs of the island'.[149] Elated, Tooworero sent these items ashore and was about to follow them when another canoe approached *Chatham* and warned him that some other chiefs on shore now felt disrespected, had seized his gifts, and were on their way to ask Kamehameha's permission to have him killed!

Archibald wrote in his journal 'We certainly little expected that while we were conferring these little acts of attention on our friend, that we were at the same time exposing him to the envy and revenge of his countrymen'.[150] But that is exactly what happened. They had to keep Tooworero on board *Chatham* overnight to allow tempers to cool. The next day he went back on shore under negotiated protection of Chief Keeaumoku.

Reflecting on the incident, Archibald may have wondered if Vancouver had wisely understood that generosity can be dangerous, and had carefully judged that providing too much wealth to someone of lower rank might upset the established social hierarchy and lead to exactly the sort of problem Tooworero faced. Or had Vancouver unwisely short-changed Tooworero and caused the whole mess to begin with? At such times, Archibald may have appreciated his status as an independent botanist reporting to Banks, rather than someone directly under Vancouver's command.

Trade and culture

Archibald would have also observed carefully the ships' various trading activities. As had been made clear with Pitt and his lashings, trade was firmly regulated by Vancouver. As botanist, Archibald's principal interest in trade was with an eye to developing potential agricultural exports from Hawaii and, on a personal level, to barter and trade for cultural artefacts – bows, arrows, cloth, knives – and to hire, reward, and thank those who helped him on his expeditions. He traded with red cloth and various items made from iron, but never with weapons. On this he and Vancouver were of one accord and they both shared concerns about the trade in weapons they had observed, and had been pressured to conduct. Commercial traders who had visited earlier were not as scrupulous on this point, and Archibald could plainly see that firearms were now more plentiful than on his previous visit aboard the *Prince of Wales*.

Visiting Chief Keeaumoku's residence with Vancouver, Archibald observed that European weapons were making the mighty even mightier. Reporting on this visit he wrote,

> Whether it was out of respect to their own chief or to Captain Vancouver I could not learn, but all the natives cowered down as we passed on through the village in this martial parade to the chief's residence, where we found his wife, mother and two sisters seated on a mat under a canopy in front of the house. … After making them some presents, consisting of beads, looking glasses and scissors, etc., and refreshing ourselves with cocoanut milk, they expressed a curiosity of seeing the marines go through their manual exercises, in which Captain Vancouver readily gratified them. The chief then showed us a large war canoe he was building, and asked Captain Vancouver to give him as much English canvas as would make a sail for it, which was promised him. He also took us into a house where he showed us several muskets that were kept in very good order, and amongst them was a double-barreled fowling piece, two swivels and a carronade. These last two were to be placed in his war canoe.[151]

But Archibald may have been equally comforted to find that trade was possible without trading in arms. One day, when anchored off Kealakekua, a few canoes approached the *Discovery* looking to trade and were satisfied simply to trade hogs for pieces of the attractive red cloth the ships had brought with them. All too often these trading encounters included a bid to trade produce for muskets and ammunition but in this instance no one even asked about it. Reflecting on this in his journal at day's end, Archibald wrote,

> It would redound more to the honour of humanity had those vessels, who had hitherto dealt with them, acted upon the same principles, by which they might still have been kept without the use of those destructive weapons that have been so industriously dispersed amongst them and which serve to stir up their minds with a desire of conquest, ruin and destruction to their fellow creatures.[152]

Gift-giving – the 'Presents' Lord Grenville had mentioned – was an important part of diplomacy, cementing friendships, showing respect, and affirming authority. In this vein Archibald would have been awestruck by the pageantry which greeted *Discovery* and *Chatham* when they first arrived at Kealakekua Bay, in February 1793. King Kamehameha pulled out all the stops and turned the event into a spectacle of pomp and power. After both ships dropped their anchors they were immediately surrounded by hundreds of canoes and Archibald observed that 'upon the most moderate computation' they were greeted by no fewer than 3,000 people, not counting those gathered on the shore.[153] King Kamehameha himself eventually arrived,

> ... standing upright in the middle of the canoe with a fan in his hand, and was gracefully robed in a beautiful long cloak of yellow feathers, and his under dress consisted of a loose gown of printed cotton girded on with a sash which he said had been given to him by Captain Cook.

Once on board, Kamehameha presented some gifts to Vancouver and then Kamehameha

> ... told Captain Vancouver in the hearing of the officers, that the feather robe he had then about him must be carefully conveyed to King George of Britannee, as it was the most valuable present he could send him, being the only one of the kind in these islands and the richest robe any of the kings of Hawaii ever wore.

He then gave 'the strictest and most solemn injunctions that it should not be put about any person's shoulders 'till it was delivered to King George'.[154]

The feather robe was indeed precious, and an appropriate gift between kings. But Kamehameha's claim that the one he presented was the only one of its kind was not entirely true. Not long after their arrival in Kealakekua Bay Archibald encountered an old friend from his previous visit to the area aboard the *Prince of Wales* – Chief Kamakeha. The chief tried to present Archibald a feathered cloak:

> After making him some little presents and renewing our former friendship, he was handing me in as a present from his canoe a long feathered cloak, which the king [Kamehameha] observing, immediately tabooed it, thinking perhaps that it would lessen the value of his own presents of the same kind to Captain Vancouver.[155]

As Archibald had seen with Vancouver and the gifting incident with Tooworero, he now saw that Kamehameha was also keenly aware of diplomatic protocol and the damage that could be caused by being too generous. It was important not to upset the growing relationship and, while he knew Archibald held high status among his peers, Kamehameha rightly judged that it would upset Vancouver to see him publicly recognised as an equal.

As keenly interested as he may have been in matters of trade – as a tool of diplomacy and a form of currency – and in the social and political effects of commercial development, Archibald was far more interested in agriculture and the power of plants to feed the hungry, heal the sick, and provide a base for commercial trade.

His botanical exploits in Hawaii were multifaceted: he was observing, cataloguing, describing, collecting, and forwarding all sorts of notes, seeds, and samples back to Banks and others via the *Daedalus,* and he was also actively importing new plants and animals. Months earlier at the Cape of Good Hope, Archibald picked up hundreds of young orange plants and protected and nurtured them in the garden hutch on *Discovery's* quarterdeck so he could distribute them in

Hawaii. Later, when they visited California, he would select cattle and sheep to bring with them and introduce these to the islanders. Before leaving England, Archibald had also received from a friend, the Scottish gardener James Lee,[156] ' … a large assortment of garden seeds to be distributed in the course of the voyage wherever they were most likely to be most useful and beneficial to mankind'[157] and 'in compliance with his humane intention' Archibald packaged these seeds up and delivered them with the orange plants wherever he thought they would help sustain a community or assist in the development of a new crop suitable for export or trade.

Archibald knew that since Hawaii was a midway point between South Asia and the west coast of the American continent, it was a natural place for trading vessels to make repairs and replenish their supply of fresh fruit and vegetables. These new crops he introduced would add variety to the fields of sweet potato, breadfruit, and sugar cane he observed on his expeditions and, if cultivated for trade to visiting ships, could increase the islanders' wealth.

Archibald visited many plantations and was much impressed with the existing agricultural scene and excited for its potential:

> Every step we advanced through these plantations became more and more interesting as we could not help admiring the manner in which the little fields on both sides of us were laid out to the greatest advantage and the perseverance and great attention of the natives in adapting to every vegetable they cultivate as far as lays in their power, its proper soil and natural situation by which their fields in general are productive of good crops that far exceed in point of perfection the produce of any civilized country within the tropics.[158]

As a chief's gardener from the Scottish Highlands, and as a botanist who knew the power of plants to heal the sick, feed the hungry, and create wealth, and as a philosopher concerned with the rights of man and the wealth of nations, Archibald was excited by the prospect of creating a slavery-free, Pacific-based alternative to the West Indies. He thought workers from the West Indies could even be resettled here and mused in his journal that

> … men of humanity, industry and experienced abilities in the exercise of their art would here in a short time be enabled to manufacture sugar and rum from luxuriant fields of cane equal if not superior to the produce of our West India plantations … that too without slavery by merely cherishing that tractable principle of industry and labour in the inhabitants, they might be gradually led on to perform every duty belonging to a plantation with the greatest ease and cheerfulness and at very little expense, which would certainly be much more satisfactory to their employers and the world at large, than if they were ground under the galling yoke of slavery which God forbid they ever should[159] … How far preferable this would be to that disgraceful mode of slavery by which we still continue to cultivate our West India Islands.[160]

Plants and people

Care for plants and people was one of Archibald's prime interests and in Hawaii he had the opportunity to do both. On the islands he documented to science a new species of mimosa (the acacia koa tree), and noted how the tall, straight, tree was used to make canoes. He identified a new species of rumex vine, a new species of vaccinium, a large and beautiful species of vicia (*Vicia menziesii*), noted the cibotium tree (*Cibotium menziesii*), some kukui trees, a species of goose, crows, a number of flowers, a shrubby geranium (*Geranium cuneatum*, another variety of which was later named *Geranium menziesii*) and a new species of raspberry. As usual, he sent seeds back to Kew Gardens regularly, made sketches, pressed cuttings, and put some samples in the hutch aboard *Discovery*.

Archibald also made acute observations on how women made cloth from tree bark and pieces of nettle:

> The inner bark being separated from the long twigs, the exterior rind was made up into small bundles and a certain quantity of a particular kind of fern, a species of *Adiantum*, mixed with it. Both were wrapped up together in the leaves of plantains or the *dracaena ferrea*, Linn. A number of these bundles being in this manner got ready, an oven is made by digging a hole in the ground where they are put, intermixed with hot stones and covered with green leaves and earth in the same manner as they dress or bake their victuals. By this heating or sweating process, the fern imparts a reddish brown colour to the bark, which is afterwards beat out into cloth.[161]

He also documented the process by which colourful feathers are harvested for regal robes:

> They do this by spreading a little of it [breadfruit tree gum] here and there on the boughs, and placing two or three berries near it which the birds are very fond of. As they perch to eat them, they are entangled with the [tree gum], but the natives are very cautious of exterminating the birds by killing all that are in this manner caught. Many of them after being stripped of their most valuable feathers are again set at liberty and run the chance of being fleeced in the same way next year.[162]

In all his travels, from village to village and from plantation to plantation, Archibald was welcomed and overwhelmed by the peaceful nature of the locals, writing that he 'seldom observed among the lower order of the people even the appearance of an angry look toward one another, far less of threats and quarrels, and that degrading practice of fighting so often observable in the ports of civilized nations.'[163] They were, he wrote, the most friendly people imaginable. These expeditions also gave Archibald the opportunity to observe religious customs and practices, and he participated in religious observances when polite to do so. Raised as a presbyterian, he was less beholden to bureaucratic orthodoxy of church than many other Christians of his era, and reflected in his journal that an Indigenous person 'worshiping his God in a gloomy forest may be equally sincere in his prayer and derive equal consolation from his religion' as the European who 'bends his knees before the rich altar and offers up his devotion in a splendid temple.'[164] Religious forms, whatever they are, he wrote, 'ought to be equally inviolable everywhere.'[165]

Although Archibald's role was mainly botanical, his skill and reputation as a surgeon and physician soon became well known, and he was constantly called upon to assist with medical matters.

While conducting his shore duties as a botanist he also become something of a travelling field doctor. One day King Keeaumoku, who had many interactions with the ships and knew

Archibald, arranged for him to visit his dying son. Paddling by canoe four or five miles north of Kealakekua, accompanied by Keeaumoku and his wife, Queen Kaahumanu, Archibald entered a cove and landed at a village belonging to Keeaumoku.

> The reason of our putting in here was the earnest request of Keeaumoku and his wife, who were desirous of my visiting a sickly son of theirs about twenty-five years of age, to administer if possible to his relief. But I was sorry to find him so far gone that it was not in my power to be of any real service to him, for he was so emaciated, low and hectic that it was not likely he would survive it many days. This deplorable state was occasioned from a wound he received about two or three months ago by a spear which entered the side of his neck, a little above the right clavicle, and took a slanting direction downward on the inside of that bone. From that wound there was now a very serious discharge. What rendered the case still more, this wound was inflicted while he and another man were throwing spears at one another in the way of exercise and diversion. Yet for this mere accident, the unfortunate man who hove the spear, being a common person, was put to death.[166]

Another time, where he was able to be of more use, he was asked to attend a young chief named Kalaikualii who had wounded himself firing a musket 'which split his hand and divided the thumb of the left hand from the forefinger, the whole length of the metacarpal bones, and carried away the ends of two middle fingers'. Archibald wrote that 'In this condition he was brought on board to me to have his hand dressed and bandaged up and he very carefully obeyed directions given to him concerning it'.[167]

Speaking of bones, Archibald was of course still alert to the direction given to him by Banks regarding such rumoured activities as cannibalism and human sacrifice. While among the Hawaiian Islands he had seen bones of defeated enemies excarnated and paraded as war trophies – indeed he even recognised a human thigh bone used as a regal sceptre – but had not seen anything to support theories of either cannibalism or human sacrifice. He had seen no evidence in Tahiti either.

Once, however, near the village of Ka'awaloa, Archibald heard a rumour of human sacrifice. A desperate-looking man had paddled out to the *Chatham* after a chief died, pleading for sanctuary. Archibald wrote in his journal,

> No sooner had the chief breathed his last than a good looking man went alongside of the *Chatham* apparently in great perturbation, and entreated in the most earnest manner to be admitted on board of her as his life was in danger of being sacrificed as a victim to the manes of the departed chief. This was at first treated as a fabricated story with intent either to impose on them or gain some end, and they were unwilling to give credit to it until Isaac Davies, who had been some years on the island, and who was then on board, informed them that it was a customary thing with these natives when any great chief died to make some human sacrifices at the same time according to the rank he bore, and that those who were pitched upon for this un-human purpose were generally the most esteemed and the most faithful adherents of the deceased.[168]

So, the man was given sanctuary on *Chatham* for three days then went back to shore when things had cooled down.

Shortly after, Archibald made enquiries as to his fate 'I made particular enquiry amongst the natives in order to learn if any other man was put to death on this occasion, but I never could receive any satisfactory information, for one person affirmed it and another denied it'.[169]

Mauna Loa

Archibald had many adventures while in Hawaii, but none was a greater high mark in his personal list of achievements than his hike to the top of the highest peak on the island: Mauna Loa.

At nearly 4,300 metres above sea level the mountain (actually a volcano) was three times higher than Ben Nevis, the highest mountain in all of Britain. Neither Archibald nor anyone else could recall seeing a higher peak in all their travels. But at this time no one knew for certain just how high it was. It was a dominant feature of the landscape, and Archibald was eager to see what plants grew at such altitude, and also keen to formally measure the height with the portable barometer and thermometer he had acquired in Cape Town.[170] He discussed the idea with Vancouver, who agreed to it; and he also discussed the expedition with King Kamehameha, who also gave his consent and assigned Chief Luhea to provide logistical support. So, on 6 February 1794 Archibald calibrated his portable barometer on *Discovery* then set off with Chief Luhea and 20 paddlers in a large double canoe belonging to the king. Baker, and two midshipmen – George McKenzie from *Discovery* and Thomas Heddington from *Chatham* – joined the party, and local resident Mr John Howell (an episcopalian clergyman, referred to as 'Padre' by the sailors) followed along in a separate double canoe with his own attendants.

They landed at Honaunau and two days later were able to leave their canoes behind at Pakini village and take the rest of the expedition by land. On the 10th they managed to hike nearly 20 miles of terrain, increasingly upwards, followed from village to village by curious crowds, with Archibald taking barometer and temperature readings at each step along the way.

Two days further on, the expedition team led by Archibald passed through a plantation that belonged to Kalaikukalii, the chief whose hand had been wounded by musket accident and whom Archibald had treated. Unbeknownst to Archibald, the chief – in gratitude for the medical services rendered – had sent word ahead that he and his party should be well-provisioned and treated when they passed through. Archibald was very touched by the gesture, noting in his journal,

> Little acts of hospitality and kindness are acceptable in all places and on all occasions, but nowhere more particularly so than to way-worn travellers in remote regions … where those little civilities may be considered the spontaneous offerings of the heart, and cannot fail to touch the feelings of those on whom they are conferred with a more than common sense of gratitude and admiration.[171]

They were now close enough to the top of Mauna Lao that they could smell and see smoke, dust, and ashes from the volcano. Archibald sent a courier back to Vancouver 'to relive any anxiety he might be under respecting us, and to acquaint him with the distance we had come and the probable time it would take us to accomplish our object'.[172] All was going well, except for one incident which may have reminded Archibald of the lost ribbon affair that had sent Vancouver into a fury. A knife was stolen at camp, and this time it was Chief Luhea who wanted the culprit to be found and punished. Archibald relayed the scene:

> One of the gentlemen laying down his knife carelessly had it stolen from him. This was made known to Luhea, who immediately caused diligent search to be made for it, and made such a stir about it amongst the whole party that it was soon found again. One of the strangers who had followed us up was suspected of having concealed it, for which the chief was in such a rage at him for this dishonesty that he certainly would have put an end to his existence on the spot by plunging his knife into his body had we not interfered at the moment he had his hand lifted over him to commit the horrid deed.[173]

Fortunately, the light-fingered follower was kicked out of the camp instead.

The night before their final ascent they camped rough:

> For our bed we made choice of a flat even rock on which we could all huddle close together, and after marking out the exact space we should occupy of it, we raised a small parapet round it with the lava to break off the wind, which after sunset blew keen and penetrating. All the bedclothes we hitherto required were a few folds of their Sandwich Islands cloth over us with a mat on the bare lava rock, as it was all agreed we should sleep together to keep ourselves warm, we joined together everything we had, for a general covering, made pillows of hard lava, and in this way passed the night tolerably comfortable, though we could not sleep much.[174]

Despite this literal hardship, Archibald was in his element. He gazed upon the grand canopy of heaven and, setting aside fears of survival, contemplated

> … the awful and extended scene around us, where the most profound stillness subsisted the whole night, not even interrupted by the least chirp of a bird or insect. The moon rose out of the sea at an immense distance, and her orb appeared very uncommonly large and brilliant, and the sky being perfectly clear overhead, the assemblage of stars appeared very numerous and shone with unusual brightness. These led the imagination to the utmost stretch and afforded objects of both wonder and admiration.[175]

The next morning they burned their sleeping mats, walking sticks, and every non-essential item they had, to make fire for heat and to cook breakfast: a small quantity of chocolate, a few ship's biscuits, a quart of rum (a little over a litre), and a few coconuts.

Those unwilling or unable to go further were sent back down, as the final party, now fewer than ten strong, gazed up with determination and awe at their target. By eleven o'clock they arrived at the summit of the volcano, at the southern end of a crater some 5 km in diameter. But they were not yet at the absolute highest point. Archibald could see that the very highest point was on the western edge of the crater, and if he was to make an accurate barometer reading, they should have to cross a large hollow full of hideous chinks and chasms of rock in all directions, strewn over with large masses of broken and peaked lava in irregular piles 'exhibiting the most rugged and disruptive appearance that can possibly be conceived'.[176] It was too much for Mr Howell. He was exhausted, his shoes were already cut to pieces, and he could go no further. He and a couple of Hawaiians decided to rest at the southern end of the crater while the final four – Archibald, Baker, McKenzie, and one of the Hawaiians – set out for the western peak. By noon they made it, and Archibald was able to take the barometer reading and calculated the height to be 13,634 feet (4,156 metres).[177]

With the barometer reading completed, the four finalists headed back across the rugged hollow of the crater, stopping two or three times for the thoroughly exhausted Lt Baker to recover his strength. Eventually they reached the place where they had left Howell but were alarmed to discover the dispirited clergyman had left:

> When we came back to the place where we left Mr. Howell and the natives, we found only two of the latter in waiting for us, faithful (poor fellows) to their trust, though shivering with the cold at the risk of their lives, and patiently enduring the pangs of both hunger and thirst.[178]

For a moment their spirits were crushed by the additional news that the departed Padre had taken their last drop of rum with him, and Archibald said the news sounded like the knell of death in their ears, adding,

> The absence of our cordial on which we had built our only hope of cheering comfort to enable us to go through the long journey still before us afflicted us most. Thus overwhelmed, spiritless and faint, we threw ourselves down on the bare rocks and for some moments revolved our melancholy situation in silence.[179]

Luckily for the summit-achievers, the two natives waiting for them near the top still had three coconuts with them and, while not as rewarding as a shot of stomach-warming rum, the milk proved to be restorative and 'after eating some of the kernels, which were carefully divided among us, we set out on our return to the encampment, where we were so fortunate as to arrive safe at ten at night, after the most persevering and hazardous struggle that can possibly be conceived'.[180]

Once back aboard *Discovery*, after two weeks on the Maua Loa expedition, Archibald was finally satisfied he had done as complete a scientific survey of the Hawaiian Islands as could be accomplished within the parameters of the overall mission. The time spent here, and his success prevailing upon Vancouver to accommodate and indulge his botanical pursuits, would gratify Banks and King George, and had been a personal and literal high. It would also add much to the knowledge of the area and inform future visits.

But, despite the literal and other heights he had reached in Hawaii, the focus of his mission lay nearly 3,000 nautical miles further east, in the Pacific Northwest. This was also where he would make his most important discoveries … and his most profound memories.

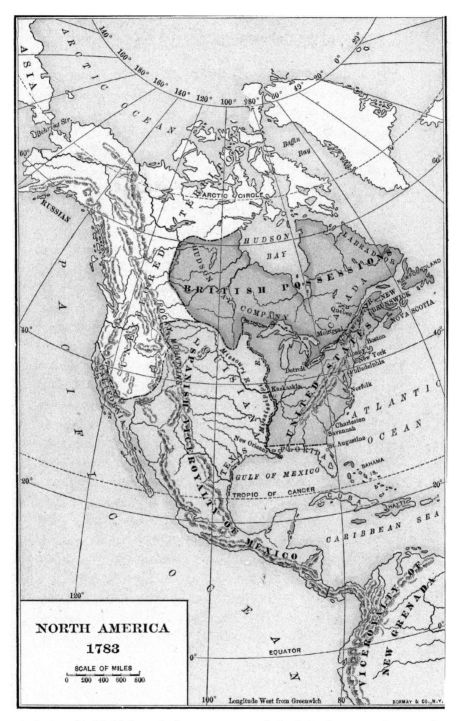

This page: At the time of Archibald's first and subsequent voyages to the Pacific Northwest coast of continental America, the area was uncharted and its indigenous communities unaligned with any European power. (North America 1783 Map from The American Nation Vol 10 (New York, NY: Harper and Brothers, 1906). *Overleaf*: L'Amérique septentrionale, ou se remarquent les États Unis, 1783 Courtesy: Library of Congress, Geography and Map Division)

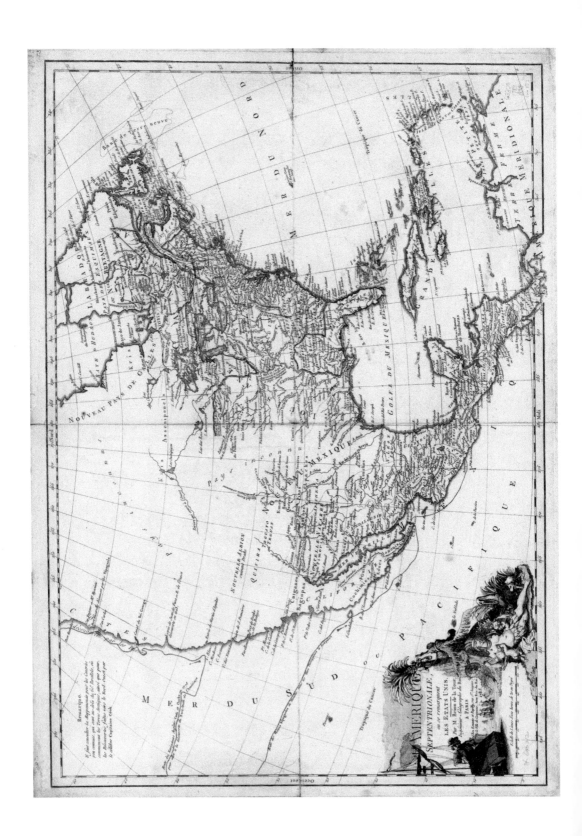

Archibald's London: 1. Archibald and Janet's last Home (Notting Hill) 2. Archibald and Janet's First Home / Medical Practice (Cavendish Square) 3. Joseph Banks' Home (Soho Square) 4. Adam Brown's Home (Brompton Square) 5. John Walker's Home (Berkeley Square) 6. Buckingham Palace 7. Location of Pitt's Duel 8. St George's Church (Hanover Square) 9. Covent Garden 10. Drury Lane Theatre 11. Osborne's Adelphi Hotel 12. Burlington House / Linnean Society 13. Thomas Richmond's Portrait Studio. (Map by author)

The ancient standing stones of Croft Morag, backdrop to Archibald's youth. (Image by author)

Above: Castle Menzies in 1748, home to the Clan Chief. Just beyond the castle are the gardens where Archibald worked as a teenager. (Painting by Paul Sandy Courtesy Scottish National Gallery of Modern Art)

Archibald's most important patrons and mentors: his Clan Chief Sir Robert Menzies (*Above left*), and the King's Gardener in Scotland, Professor John Hope (red cape, *Above right*). (Painting attributed to Sir John Baptiste de Medina, image by author; Illustration by John Kay, public domain)

King George III as "Farmer George" (*Left*); Archibald's third mentor and patron Joseph Banks (*Right*), in a portrait painted after his voyage with Cook. (Farmer George image © Trustees of the British Museum; Painting by Benjamin West Courtesy the Usher Gallery / Bridgeman)

The Battle of the Saintes was a vicious yet decisive victory for the Royal Navy, crushing the French forces and returning naval dominance of the east coast of the newly independent United States to Great Britain. (Painting by François Aimé Louis Dumoulin at Musée Historique de Vevey)

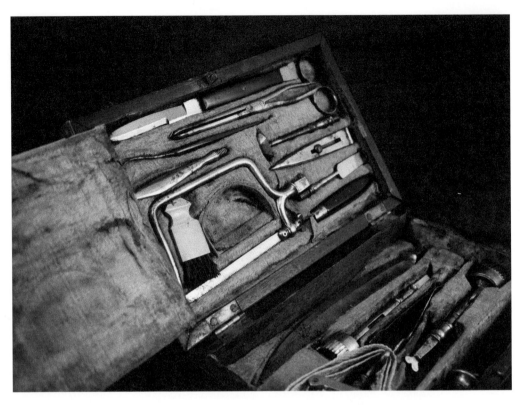

Tools of the trade: A Royal Navy surgeon's kit (*above*), and a botanist's vasculum (*below*). (Image courtesy Royal College of Physicians and Surgeons of Glasgow (ref 2019/6; Image by Ji-Elle at Musée botanique de Berlin)

Left: Chief Maquinna, 1791. "Maquinna together with his Brother and Attendants received us on the Beach, and we were conducted to the Chief's House which was large and spacious and occupied by himself, his Brother and other families of distinction." [Archibald's Journal Sept. 5, 1792] *Below*: A Kwakwaka'wakw mask. "a young man, appearing to be the chief of the party, seated himself in the bow of the yawl, and put on a mask resembling a wolf's face" [Vancouver's Journal, August, 1793] (Illustration by Tomås de Suria at Museo de América; Mask image courtesy UBC Museum of Anthropology)

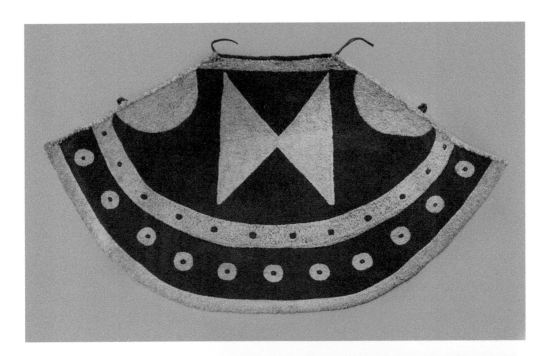

Above: Chief Kamakeha presented Archibald a cloak of feathers: "After making him some little presents and renewing our former friendship, he was handing me in as a present from his canoe a long feathered cloak, which the king [Kamehameha] observing, immediately tabooed it, thinking perhaps that it would lessen the value of his own presents of the same kind to Captain Vancouver" [Archibald's Journal, March 5, 1793]
Right: A Royal Sceptre made from a human thigh bone, collected through trade by Archibald or presented to him as a gift. (Both images © The Trustees of the British Museum)

The Canton Factories, where the King George's Sound Company sold its Pacific Northwest furs before the Prince of Wales – under command of Johnstone and with Archibald and Hawaiian lad Toowerorero - sailed back to London loaded with silk, china, and other valuable trade goods. (Painting by William Daniel at National Maritime Museum, Greenwich, London)

HMS Discovery arrives at Nootka Sound. "There were at this time ten Vessels riding at Anchor in this small Cove, besides two small ones building on shore … and this perhaps was the greatest number of Vessels hitherto collected together in this Sound at any one period." [Archibald's journal, Sept. 22, 1792]. (Painting by John Horton, courtesy John Horton)

VISTA INTERIOR DE LA CASA DE MACUINA, EN QUE SE REPRESENTA ESTE JEFE BAYLANDO, Y SUS DOMESTICOS CANTANDO Y TOCANDO.

Chief Maquinna hosted British and Spanish guests at a potlatch ceremony. "a Throne was erected on which the young Princess was seated by her Father, and from thence Copper Iron Beads etc. and every other article of any value the Chief possessed was thrown down and scattered in the most profuse manner amongst the people" [Archibald's Journal, Sept 5, 1792] (Illustrator unattributed; "Macuina's House", Ministry of Foreign Affairs, Madrid, Spain)

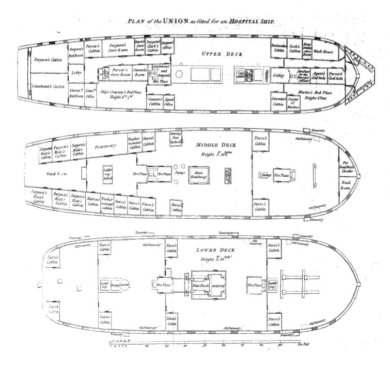

Floorplan of the Union Hospital Ship, where Archibald conducted viral epidemic fumigation experiments. "the vapour having entirely subsided, the ports and scuttles were thrown open, for the admission of fresh air. I then walked through the wards, and plainly perceived that the air of the hospital was greatly sweetened, even by this first fumigation." [Archibald's notes November 26, 1795] (Illustrator unknown, public domain)

Left: The newlyweds' miniature portraits, created by Thomas Richmond at his Green Park studio. Archibald (age 48) wearing a blue velvet frock coat and his light brown hair powdered. Janet (age 32) wears berry-red earrings and a loosely flowing garment reminiscent of the robes of classical philosophers. *Right*: The ladies' fashion of the day. (Image PDP05373 and PDP05374 Courtesy of the BC Archives; illustration from Lady's Monthly Magazine, Sept 1798 at the Los Angeles County Museum of Art)

The original bust of Archibald (*Left*) by Canadian sculptor Jack Harman in 1989, and its twin (*Right*) cast by Stephen Harman (*Centre*) in 2020. *Right*: The Archibald Menzies Room at Castle Menzies. The window at left has a view toward the gardens he worked as a boy. (Image courtesy Stephen Harman; image by author)

Eight
WONDER IN THE WEST COAST (SUMMERS 1792–1794)

The winter months in Hawaii had delivered many highs, but the principal focus of the Vancouver mission was located all along the Pacific Northwest coast of America, and here Archibald spent most of his time while assigned to the *Discovery*. In all, he spent about 24 months, between the Spring of 1792 and summer of 1794, roaming between California and Alaska. Most of those months would be in waters off what is known today as the coast of British Columbia.

'Columbia' is an important part of the place name 'British Columbia' and an often-overlooked clue to the province's origins. The word refers to the Columbia River, located about 300 kilometres south of the current Canada–US border. At the time of the *Discovery*'s visit to the West Coast, the Columbia River was just another vaguely known geographical feature of the ambiguous, mostly uncharted, territory which lay north of the Spanish settlement in San Francisco and far south of Russian interests in Alaska. But, as later years would prove, it was one of the geopolitically important places the *Discovery* visited while looking for the elusive Northwest Passage.

The Columbia River was also elusive – the coastal area near it was so poorly charted it took *Discovery* two attempts to find it.

The first attempt, if indeed it can truly be called that, was a few days after Archibald and the crew first set eyes on the West Coast, about a month after leaving Hawaii. They sighted land, near Mendocino, California on 17 April 1792 and began to make their journey north from there, not landing but charting and surveying the coastline from the deck of the ship. There was always the possibility of sighting another ship in the area, but sometimes their eyes played tricks on them. Archibald wrote in his journal that he 'Saw some Whales, the spoutings of these afterwards in the hazy horizon loomed so as to be taken for strange vessels under sail, and it was even some time before the deception was clearly detected'.[181]

As the expedition sailed north towards their intended destination, they came to the area where the Columbia River poured into the Pacific but didn't stop to investigate further. 'The sea has now changed from its natural, to river coloured water, the probable consequence of some streams falling into the bay, or into the ocean to the north of it, through the low land' wrote Vancouver, adding 'Not considering this opening worthy of more attention, I continued our pursuit to the Northwest'.[182]

The second and more fruitful attempt was in October of the same year, when reversing their route and preparing to spend the winter months back in Hawaii. This time Vancouver hung around the mouth of the Columbia and, fearing the bar at the head of the river too risky for *Discovery* to pass, sent *Chatham* to explore further.

Vancouver was not merely curious; he was pursuing one of the mission's key objectives: to discover a passage to the Atlantic or, failing that, a connection between the Pacific coast and the known lakes in the interior of British North America. As the Admiralty had commanded them, in the somewhat rambling style of the day:

> It would be of great importance if it could be found that, by means of any considerable inlets of the sea, or even of large rivers, communicating with the lakes in the interior of the continent, such an intercourse, as has been already mentioned, could be established; it will therefore be necessary, for the purpose of ascertaining this point, that the survey could be so conducted, as not only to ascertain the general line of the sea coast, but also the direction and extent of all such considerable inlets, whether made by arms of the sea, or by the mouths of large rivers, as may be likely to lead to, or facilitate, such communication as is above described.[183]

Using slightly more direct language, the Admiralty said Vancouver should acquire

> … accurate information with respect to the nature and extent of any water-communication which may tend, in any considerable degree, to facilitate an intercourse, for the purposes of commerce, between the north-west coast, and the country upon the opposite fide of the continent, which are inhabited or occupied by His Majesty's subjects.[184]

The crew relentlessly pursued this goal. Broughton successfully navigated the smaller *Chatham* over the sandbar at the mouth of the river, then took a team on row boats about 160 km further up the river. He did not find a path to Hudson Bay, the Great Lakes, or the Atlantic, but he did chart the river and, though he may not have been fully aware of it at the time, this feat alone would play an important role in later territorial negotiations with the ever-expanding American republic, recently established on the east coast.[185]

The more promising body of water investigated by the *Discovery* and *Chatham* on their quest for the Northwest Passage was the Strait of Juan de Fuca. Like the Columbia River, the strait was, at the time of Archibald's arrival on the scene, still something of a mystery. Captain Cook had heard about it but dismissed it as a Spanish fantasy. Yet it was a powerful fantasy that had persisted over the years since Cook left, and even a remote chance that it was the key to the Northwest Passage was too important not to investigate thoroughly.

When Archibald was first in the area with Colnett, they sailed right past the strait, as Cook had done, and touched land directly only at Nootka, which (like everyone else) they believed to be the West Coast of continental North America. But Colnett's was a commercial mission; now they were directly charged with finding and investigating the Strait. Indeed, the Admiralty specifically required and directed them to

> … pay a particular attention to the examination of the supposed Straits of Juan de Fuca, said to be situated between 48 and 49 north latitude, and to lead to an opening through which the sloop *Washington* is reported [by Colnett's outspoken business partner John Meares] to have sailed in 1789, and to have come out again to the northward of Nootka.

The discovery of a near communication between any such sea or strait, and any river running into, or from the lake of the woods, would be particularly useful.[186]

Exploration and discovery

Remarkably, just as *Discovery* and *Chatham* first approached the entrance to the Straits of Juan de Fuca, two hundred kilometres north of the Columbia River, in April of 1792, they spied an American trade ship Captained by Robert Gray … the very same man who had previously been captain of the sloop *Washington*, mentioned in their orders.

Vancouver sent Archibald and Puget over to Gray's ship, via row boat, to find out what they could, and they enthusiastically drained him of whatever information he was prepared to share. Archibald and Puget returned with the news that, according to Gray, the Strait of Juan de Fuca did exist – indeed they were basically at its mouth now – but he had also laughed at Meares' claim that he had sailed through the Straits to come out north of Nootka.

> Mr. Gray informed us that in his former Voyage he had gone up the Straits of Juan de Fuca in the Sloop *Washington* about 17 leagues [about 95 kilometres] in an East by South direction and finding he did not meet with encouragement as a Trader to pursue it further he returned back and came out to Sea again the very same way he had entered – he was therefore struck with astonishment when we informed him of the sweeping tract of several degrees which Mr. Mears had given him credit for in his Chart and publication.[187]

With this intelligence secured, Vancouver determined to thoroughly explore and chart the Straits. He continued sailing east, then south-easter as Gray had advised, until he entered an area which he subsequently named Puget's Sound. They then spent the next several months exploring and charting the entire area, probing every nook and cranny of the coastline, seeking the Northwest Passage. As it was difficult to distinguish the continental coastline itself in an area full of islands, Vancouver soon established a routine of sailing the two ships to a safe anchorage, then dispatching the small boats to conduct individual sorties among the numerous inlets and passages. Johnstone, Whidbey, and Vancouver commanded the small boats, and Archibald took turns travelling with each of them on different botanical expeditions.

Though small compared to the ships (at about 6 metres long), each boat had was big enough to carry a dozen crew plus camping equipment and supplies to last for a week or more. Food could be foraged, hunted, or traded for among the Indigenous people they encountered. Vancouver set an aggressively strong pace – in one three-week period the team rowed over 1,300 km to chart 97 km of coastline as the officers made their navigational charts and Archibald made his botanical survey. On one particularly rigorous day Vancouver drove his small boat crew to row for a total of 18 hours.

Archibald enthusiastically recorded his experiences around Puget's Sound, peppering his botanical notes with keen observations of the people he met. At Eld Inlet, near present-day Olympia, Washington, Archibald estimated the population of the two villages at between 70 and 80 people – 'We visited one of [the villages] and they received us in the most friendly manner without shewing the least signs of distrustful behaviour or any fear or alarm when we landed amongst them' – and noted how their huts were covered with mats made from bulrushes:

… and their Dresses were chiefly formed of the Skins of the wild animals of the forest, that which was peculiar to the Women was a dressed Deer's Skin wrapped round their waist and covering down to their knees or rather below them, and the men too generally wore some little covering before them to hide those parts which modesty and almost the universal voice of nature require. We made but a short stay among these people and on leaving them distributed some Beads and little ornamental Trinkets chiefly of Brass and Copper among the Women and Children of which they were very fond.[188]

A few days later, near present-day Vashon Island, Archibald was able to document some new plants thanks to the industriousness of the local women at another community of about 70 people. Landing his small boat on a point near where *Discovery* was anchored, he noted,

Several of the women were digging on the Point which excited my curiosity to know what they were digging for and found it to be a little bulbous root of a liliaceous plant which on searching about for the flower of it I discovered to be a new Genus of the Triandria monogina".

Enquiring further he learned that the root with young shoots of raspberries, and a species of barnacle harvested from the rocks along the shore, formed a chief part of the local diet. He also humorously admonished that while the women were occupied making mats of the bulrushes 'the men were lolling about in sluggish idleness'.[189]

One of the most remarkable (and most obvious) landscape features in the Puget's Sound area was a large volcanic mountain peak in the distance. Archibald, ever the Highland hillclimber, eyed the mountain respectfully and wrote in his journal,

In the North East quarter a very high solid ridge of Mountains was observed one of which was seen wholly covered with Snow and with a lofty summit over topping all the others around it upwards of twenty leagues [97km] off nearly in a North East direction— This obtained the name of *Mount Baker* after the Gentleman who first observed it.[190]

It was an impressive sight, its peak dominating the other features nearby, but too far inland for Archibald to climb with his portable barometer.[191]

As they rigorously charted and mapped the area, Vancouver continued to assign names to prominent places. As was the fashion of the times, naval officers' surnames were a common source of inspiration for him. A month after naming Mount Baker they passed a 'fine large island' which Vancouver called Whidbey Island. After having thoroughly botanised and charted the southern extremes of the Puget's Sound area, they travelled north along the west coast of the continent, passing another navigational landmark which Vancouver named Point Roberts in honour of *Discovery*'s intended first captain, Henry Roberts. Not much farther along the coast he named Point Grey for his friend Captain George Grey who, like Vancouver, began his naval career at age fourteen; Archibald knew Grey as a fellow veteran of the Battle of the Saintes. Observing a large channel further to the east, Vancouver assigned it the name Burrard's Channel (now Burrard Inlet) after his naval friend Harry Burrard, just twenty-six years old at the time but a combat veteran destined to become 'Sir Harry' and a senior officer of the Admiralty.

Leaving Burrard's Channel they saw to the north another promising pathway worth exploring but which proved to be a dead-end. On his ever-expanding charts, Vancouver named this Howe Sound after Richard Howe, the recently retired first lord of the admiralty and veteran of many naval battles, notably in the American Revolutionary War. Pressing on, they soon found another grand route that opened up to the north which, in honour of His Majesty, Vancouver named the Strait of Georgia.

While Vancouver used the names of naval officers for inspiration, Archibald was more inspired by the Linnaean taxonomy system. He called one tree the *acer circinatum* (vine maple), another he identified as *alnus incana rubra* (red alder), to a berry plant he assigned the name *amelanchier alnifolia* (saskatoon berry) and to a flower the name *juniperus scopulorum* (Rocky Mountain juniper). But on some occasions his botanical exploits provided the backstory behind a few of the places Vancouver named. Birch Bay, just a few miles south of Point Roberts was one such instance. Archibald's journal entry for 12 June 1792 explains how his enthusiastic botanical reports inspired Vancouver to expand his naming repertoire:

> I landed at the place where the Tents were erected and walked from thence round the bottom of the Bay to examine the natural productions of the Country and found that besides the Pines already enumerated the Woods here abounded with the white and trembling Poplars together with black Birch. In consequence of my discovery of the latter place, the place afterwards obtained the name of *Birch Bay*.[192]

Another time, in the Desolation Sound areas (so named by Vancouver in consideration of the few resources he found there),[193] while anchored near Redonda Island, Archibald spied another hill to climb. Taking his trusty barometer and vasculum, he persuaded Mudge and a band of hearty sailors to join him on the hike:

> In order to vary my excursions and search the upper regions of the Woods for Botanical acquisitions I one day ascended a hill on the North Side of the Channel close to the Ship in company with some of the Gentlemen, and found my journey amply repaid by a number of new Plants never before discovered. As we did not know the time it might take us to reach the summit, we took with us some men to carry provision and water and landed pretty early in order to have the fatiguing part of the Journey over before the heat of the day.[194]

He added enthusiastically,

> In this journey the Genus Pyrola was enriched with four new species which I met with nowhere else and on the top of the Hill I found two new species of Pentstemon, a new species of Ribes Andromeda coerulea, Pinus Strobus—Pinus inops. H.K. and a great variety of Cryptogamic Plants, besides many other undescribed plants which I had before met with in other parts of the Country.[195]

Nearby he also noted,

There was a beautiful Waterfall which issued from a Lake close behind it and precipitated a wide foaming stream into the Sea over a shelving rocky precipice of about thirty yards [27 metres] high, its wild romantic appearance aided by its rugged situation and the gloomy forests which surrounded it, rendered it a place of resort for small parties to visit during our stay.[196] On the Banks of this Lake I found the following Plants. Linnoea borealis, Myrica gale, Anthericum Calyculatum, Drosera rotundifolia, Menyanthes trifoliata, Shanus albus, and in the Lake itself we found some Bivalve Shells which were quite new to me.[197]

The climb to the top of the hill paid off not only with these many new plants Archibald was able to name and classify for the benefit of the king's gardens at Kew, but also with the panoramic southerly views which helped Mudge chart the area. This led Vancouver to name a nearby landmark Cape Mudge.

Not long after this hike Archibald and Mudge persuaded Vancouver to visit the village located there. Archibald described a steep elevated bank with a village of about 12 houses planked over with large boards, some ornamented with paintings. He could not determine the precise meaning of the paintings, but they reminded him of the painted armorial crest carved in stone over the entrance to his own clan chief's home and supposed they might also be expressions of clan identity or legend. The wooden structures reminded him of the stone black houses he had seen in the Hebrides, and he understood their purpose was much the same. Taking in the entire village he described homes that

> ... were flat roofed and of a quadrangular figure and each house contained several families to the number of about three-hundred-and-fifty Inhabitants in all on the most moderate calculation, for there were eighteen Canoes alongside of the Ship before we left it, and on landing we counted about seventy on the Beach, so if we allow only four persons to each Canoe which is very moderate it will give upwards of the number we have above computed.[198]

Archibald described the people, undoubtedly of the We Wai Kai nation, as being of average height, slender, and of a light copper colour:

> They have flat broad faces ... their Ears are perforated for appending Ornaments either of Copper or pearly Shells; the Septum of the Nose they also pierce and sometimes wear a quill or piece of tooth-shell in it; their Hair is streight black and long, but mixed with such quantity of red-ochre grease and dirt puffed over at times with white down that its real colour is not easily distinguishable ...

He added that some had ornamented their faces with red-ochre paint, sprinkled over with black glimmer 'that helped not a little to heighten their ferocious appearance'. He went on to note the women and children did not appear in any way shy or timorous, although he felt sure their party were the first Europeans they had ever seen. The women were dressed with 'Garments made either of the Skins of wild Animals or wove from Wool or the prepared bark of the American Arber Vitae Tree [northern white cedar]' but 'many of the Men went entirely naked without giving the least offence to the other Sex or shewing any apparent shame at their situation.'[199]

While Vancouver pored over his navigational charts and plotted whether to continue searching for the Northwest Passage or turn south and head back out of the Straits of Juan de Fuca, to approach Nootka from the south, Archibald, in his friendly and curious manner, lingered on land and continued to make observations about the technology and materials used in the construction of hunting and fishing tools, and he traded for some artefacts he could send back to fellow naturalists and ethnologists to study in London via the *Daedalus*.

He wrote,

> Their Fish-hooks are nearly the same as at Nootka Sound and we saw some Fishing-Nets drying upon stakes before the houses; their Bows were lined with Sinews and shaped like those we saw on the East Side of this great Gulph, and the Arrows were also fastened in the same manner, but most of them were armed with pieces of Muscle Shell instead of flinty stones. Their Canoes were small with projecting prows and dug out of one piece of Timber each with four or five small thorts [?]and some of them had their outside ornamented with rude [rough] figures painted with red-ochre: their Paddles were short with round handles and pointed blades. Some Fish and Curiosities were purchased from them for Beads and small Trinkets, and in these little dealings they appeared to be guided by the strictest honesty, indeed their whole conduct during our short stay was quiet [*sic*] friendly and hospitable, pressing us often to partake of their entertainment such as Fish Berries and Water, and we in return endeavoured to make them sensible of our approbation by distributing among the Women and Children some small presents, which made them appear highly gratified.[200]

Walking some short distance from the village, Archibald observed local burial traditions and speculated whether the custom was for everyone or just for chiefs and families of distinction. Walking westward along the side of the channel on a pleasant, clear, level pasture for nearly 3 km, he

> … observed in the verge of the wood their manner of disposing of their dead which was by putting them either in small square boxes or wrapping them well up in Mats or old garments into square bundles and placing them above ground in small Tombs erected for the purpose and closely boarded on every side.[201]

He had made similar observations at Birch Bay:

> In one place in the verge of the Wood I saw an old Canoe suspended five or six feet from the ground between two Trees and containing some decayed human bones wrapped up in Mats and carefully covered over with Boards; as something of the same kind was seen in three or four instances to the Southward of this, it would appear that this is the general mode of entombing their dead …

Though he could only speculate why this method was used: ' … what gave rise to so singular a custom I am at a loss to determine, unless it is to place them out of the reach of Bears Wolves and other Animals and prevent them from digging up or offering any violence to recent bodies after interment'.[202]

While Archibald was exploring the area around Cape Mudge, his friend Johnstone had been probing the waterways further to the north in one of the small boats and made a major navigational breakthrough: he found the passage linking Desolation Sound with the open Pacific to the north, thereby proving that Nootka was on the west coast of an island, not the west coast of the continent. Returning from his successful probe of the waterways, Johnstone immediately told Archibald of his discovery and Archibald dutifully, yet with excited haste, relayed the adventure in his journal, writing, 'Weather was frequently thick and unfavorable with heavy rain and a strong breeze of wind often against them yet they anxiously persevered in their pursuit with toilsome labour and gained sight of the Sea on the tenth of July seven days after their departure ... '[203]

On receiving this ground-breaking news, Vancouver named the route Johnstone's Straits. The Northwest Passage had not been found, but the search was yielding other significant results: many previously uncharted parts of the coast were being mapped, and the true west coast of the continent in this area was now confirmed. It was a major discovery.

Nootka diplomacy

Of course, Vancouver's other main responsibility concerned the Nootka dispute with Spain: he had to formally receive back from Spain what had been wrongfully taken from British traders three years previously. Thanks to Johnstone, they now proceeded to Nootka by sailing north up Johnstone's Strait, then followed the coast to their west until it curved down to the south.

When they sailed into Friendly Cove, on 28 August 1792, the Spanish were dutifully waiting. The gruff and aggressive Martinéz, who had sparked the crisis, was long gone, replaced by the fashionable, elegant, gregarious, forty-nine-year-old Juan Francisco de la Bodega y Quadra, commander of the Spanish naval forces along the Pacific coast of North America. Quadra had come from his base in San Blas, Mexico and had been at Nootka since March. Archibald was no doubt impressed by the changes to the port in the four years since he had last been here, and since the Spanish had arrested Colnett and triggered Meares's outrage. They had clearly made themselves comfortable. On the shore a large Spanish flag was flying high for all to see, and Quadra had deployed his blacksmiths and carpenters to build and enhance the evolving domestic infrastructure (chicken pens, a fenced-off vegetable garden, houses, barracks), and had constructed a magnificent mansion – all things considered – complete with European-styled drawing room, a balcony off the upper front level, servants' hall and kitchen, guard room, and a dining hall where Quadra, his officers and distinguished guests could dine off his silver plates and drink madeira from his crystal goblets.

Archibald likely thought the ships in Friendly Cove were as bedazzling as Quadra's dinner table: in addition to Quadra's ship several others were present including the *Sutil*, under the command of Captain Galiano, and the *Mexicana*, commanded by Captain Valdes, each proudly flying the Spanish colours. The *Discovery* saluted them with 13 guns and the honour was returned.[204] The cove which Archibald had been among the first to visit in 1787 was now becoming a popular destination, and he noted, 'There were at this time ten Vessels riding at Anchor in this small Cove, besides two small ones building on shore ... and this perhaps was the greatest number of Vessels hitherto collected together in this Sound at any one period'.[205]

After some additional time getting to know each other and exchanging diplomatic notes, Vancouver and Quadra decided they should leave their ships anchored at Nootka and, in their small boats (two from *Discovery*, one from *Chatham*, and a large Spanish launch) make for the nearby village of Tahsis[206] to confer with Chief Maquinna, whom Archibald had befriended earlier and knew was the main political force in the region.

Archibald went with Vancouver and, arriving near Tahsis towards the end of the day, the Europeans camped overnight on shore so they could make a grand entrance the following morning. Archibald wrote that after breakfast,

> We all embarked in the Boats and made a kind of martial parade with our little musical Band before the Village of *Tashees*, where we landed amidst the noisy acclamations of the Natives. Maquinna together with his Brother and Attendants received us on the Beach, and we were conducted to the Chief's House which was large and spacious and occupied by himself, his Brother and other families of distinction. Here we found the Women decently seated on Mats spread on little risings on each side of the House and Benches were placed at one end covered over with rich Furs and clean Mats for the party to set down on. We first advanced to the Royal Mat to pay our respects to the Chiefs Wives and Daughter, the latter was a young Girl about thirteen years of age named *Apinnas*, who the Spaniards informed us had been lately recognised and inaugurated in a most pompous and solemn manner by the whole Tribe as the Successor of her Father.

Archibald then witnessed, and became among the very first Europeans to describe, a potlach ceremony:

> When the Natives were assembled on this occasion, a Throne was erected on which the young Princess was seated by her Father, and from thence Copper Iron Beads etc. and every other article of any value the Chief possessed was thrown down and scattered in the most profuse manner amongst the people, who scrambled for it and expressed their approbation by continual plaudits. After this ceremony they continued their rejoicing by feasting singing and dancing for some days, till the Chief with respect to riches was brought almost upon a level with the poorest of his Tribe.

It was not the sort of extreme generosity he'd ever imagine Sir Robert or any other Highland chief exhibiting, yet he understood the ideals of *dùthchas* and *oighreachd* were not that different from what he was seeing played out here.

At the potlach he was also delighted to recognise the wife of Maquinna's brother, who had been so helpful to him when he was botanising the area years earlier with Colnett:

> On turning to those seated on the other side of the house I instantly recognized in the Wife of Maquinna's Brother an old acquaintance the daughter of an elderly chief who had a numerous family and lived in the North East corner of the Sound and to whose friendship I owed much civility and kindness when I was here about five years ago. She and her Sisters were then very young, yet they frequently, shewed so much solicitude for my safety, that they often warned me in the most earnest manner of the dangers to which my Botanical rambles in the Woods exposed me, and when they found me inattentive

to their entreaties, they would then watch the avenue of the Forest where I entered, to prevent my receiving any insult or ill-usage from their Countrymen. But it was not till after I left them that I became sensible how much I owed to their disinterested zeal for my welfare by knowing more of the treacheries and stratagems of the Natives on other parts of the Coast. I emptied my pockets of all the little Trinkets they contained in her lap and begged her to come on board the Vessel with her Father who she told me was still alive, that I might have an opportunity of renewing our friendship by some gratifying present.

These ceremonies and reunions being over with, the Spanish turned to the business at hand. Captain Quadra explained to Maquinna that the Spanish would leave and the British would stay:

> As soon as the Party was seated Senior Quadra explained to the Chief the purport of our visit and with a disinterested zeal which marked his benign character he said every thing in recommending Captain Vancouver, Mr Broughton together with their Officers and the English Nation in general to his kind attention and to a friendly intercourse with all his tribe; he assured him of the friendship and good understanding which subsisted between the English and Spaniards, and that the latter were only to quit his Territories by a mutual agreement between the two Nations.[207]

In a private moment, the two warrior-diplomats agreed that assigning their names to the entire island would be an appropriate gesture to signify and cement the goodwill that existed between them. On this matter Archibald had no right to comment and had the good sense to keep his mouth shut. But that did not stop him from later expressing his opinion in a cheeky note to Banks, writing that 'our commander has already perpetuated his name on the coast for the great island we circumnavigated last summer, of which Nootka is a part', adding that it was 'modestly named Quadra and Vancouver's Island'.[208]

While Vancouver and Quadra were not able to agree precisely where the Spanish claims ended – largely because neither of them had received clear instructions or details on this point – they were able to agree that the Spanish influence in the area would generally ramp down, while the British could ramp up unrestricted. To that extent, they were able to fulfil at least the spirit of their orders and defuse the crisis which had afflicted their two countries. Any animosity that had existed between the British and Spanish was now in the past. There would be no war over Nootka or the Pacific Northwest. Archibald summed up the feelings on behalf of the officers and himself:

> It was but natural to feel some reluctance at parting as during our stay at Nootka the Spanish Officers and we lived on the most amicable footing. Our frequent and social meetings at Senior Quadra's hospitable mansion afforded constant opportunity of testifying our mutual regard and friendship for each other, by that harmony and good understanding which always marked our convivial hours. In short ... we passed our time together chearfully and happy.[209]

Natural wonders

Archibald's remaining time in the Pacific Northwest, punctuated by winter months in Hawaii, consisted of additional voyages up and down the West Coast, from Alaska to California, investigating every nook and cranny of the rugged and irregular coastline. Vancouver, constantly on the lookout for the Northwest Passage and as always making meticulous navigational charts; Archibald constantly gathering seeds, cataloguing new species, and observing the different languages, customs, technologies, and behaviours of the thousands of Indigenous peoples scattered along the coast.

Where Vancouver often saw a harsh 'desolate' environment Archibald found a vibrant and exciting world full of natural wonders. He documented and collected several hundred plant species – 250 different terrestrial plants species in the first four months alone[210] – yet even though he was absorbed in the Latin taxonomy of Carl Linnaeus there were times when the geography of the area inspired him to write like a poet. Consider this glorious journal entry from 6 June 1792, following an off-ship romp around Admiralty Sound:

> A Traveller wandering over these unfrequented Plains is regaled with a salubrious and vivifying air impregnated with the balsamic fragrance of the surrounding Pinery, while his mind is eagerly occupied every moment on new objects and his senses rivetted on the enchanting variety of the surrounding scenery where the softer beauties of Landscape are harmoniously blended in majestic grandeur with the wild and romantic to form an interesting and picturesque prospect on every side.[211]

And again, a little over a week later (23 June), now somewhat further north at Jervis Inlet:[212]

> In going up this Arm they here and there passed immense Cascades rushing from the Summits of high precipices and dashing headlong down Chasms against projecting Rocks and Cliffs with a furious wildness that beggared all description. Curiosity led them to approach one of the largest where it poured its foaming ponderous stream over high rugged Cliffs and precipices into the fretted Sea with such stunning noise and rapidity of motion that they could not look up to its source without being affected with giddiness nor contemplate its romantic wildness without a mixture of awe and admiration.[213]

Reading these lines and others from his journal it's hard not to think Archibald was inspired by the works of Robert Burns. Perhaps he had been touched by Burns's *Birks of Aberfeldy* (with its 'foamy stream and cliffs overhung with fragrant shaws'), written near a waterfall 3 km south of the castle gardens where Archibald had laboured as a child. The natural rugged beauty of the Pacific west coast was clearly affecting Archibald much as the rugged beauty of the Highlands had affected and inspired Burns.

After two summers in the area Archibald was mesmerised by its beauty. Some people – like Captain Bligh's unruly mutineers – would surely have preferred the tropical beaches and the nubile beauties of Hawaii and Tahiti – but Archibald was smitten by the West Coast's wonders. Here, among the ponderous foaming streams and majestic grandeur of the Pacific Northwest, he felt most alive. A few months before leaving the area for the last time, he dashed off a quick note to Banks, via *Daedalus*, indicating his desire to return in a more permanent way. With Nootka relieved from Spanish control, and the British presence more dominant, he speculated the government might make further investments in its economic development. Perhaps one day it would be as big and important a port as Halifax. And a naval port of that stature would need a qualified and experienced naval surgeon. To Banks he wrote, 'If it is the intention of the government to establish a settlement on this coast … and you think it an eligible plan for me to be surgeon of it, I would esteem it a particular favour if you procure that situation for me …'[214]

Meantime he rowed, sailed, climbed, crawled, walked, and scrambled along the sandy shores, the rocky boulders, the brambles, bushes, and trees taking every opportunity to jot down his various discoveries. He even stood below, and then dissected, a wasp's nest to make his observations.

Medical man

While he had hoped he would be able to focus exclusively on botanical pursuits, Archibald was inevitably pulled into medical service. This was due partly to his aptitude, and partly to circumstance: when the mission's lead surgeon, Dr Cranstoun, took ill 18 months into the voyage, Vancouver formally asked Archibald to take over.[215] Archibald wrote to Banks to explain the situation:

> The Surgeon of the *Discovery* is to return to England in the Store Ship by the way of Botany Bay on account of the ill state of His health and Capt. Vancouver's earnest solicitations has induced me to accept his place with this proviso that he will take care it will not interfere but as little as possible with my other pursuits – indeed I have in some measure attended the Surgeon's duty since we left the Cape of Good Hope on account of Mr. Cranston's indisposition and constantly prescribed for Capt. Vancouver himself since we left England, – so that the difference now of attending the duty wholly will be very little as I have two assistants and the ship in general Healthy – besides I have by this change, got an additional Cabin which will be very serviceable in preserving my Collection, so that I trust it will meet with your Approbation, as I can assure you, that my endeavours, will suffer no Abatement, in consequence thereof in executing the object of my mission.[216]

For the most part his medical duties were routine. He attended to crew hurt in accidents, as when the *Discovery* became stuck on some rocks,[217] and a crewman was injured on the rigging: 'the forenoon was chiefly employed in re-stowing the Booms and getting up the Yards and Topmasts, on which occasion John Turner one of the Seamen had the misfortune to have his right arm fractured by the Mast Rope being carried away in swaying up the Main top gallant Mast'.[218] On another occasion he attended to one of the Indigenous men who had had a firearm accident – 'a Chief who had his Arm blown up with Gun powder'.[219] At Nootka he attended to Señor Quadra: 'He put himself under my care as a patient on our arrival here for a severe Headache of which he complained, he said, for upwards of two years, and I was extremely happy that my endeavours proved serviceable in the reestablishment of his health before he went away'.[220]

It was also easy for him to make sure the crew were kept healthy through their diet. For men living at sea for years, with no regular access to farms or markets, supplementing their diet of fish and meat with fresh vegetables was critical to their health. In this regard they could have had no better companion than Archibald. His journals are full of references to the many ways he helped the crew stay healthy: in Kitimat Arm he directed the crew to pick and boil the leaves of the beach lovage plant (*ligusticum scoticum*), and boil the entire plant of American pickleweed (*salicornia virginica*). At Tahsis he learned from the Nuu-cha-nulth women that a type of clover (a new species of trifolium) could be mixed with oil and served as a salad. When he found the *rhododendron groenlandicum* plant he advised how it could be made into tea 'a drink that was very palatable and salubrious' and 'not inferior in those qualities to the manufactured teas of China'.[221] In addition to tea, he also made sure the crew's diet included spruce beer – high in the as-yet undiscovered vitamin C – made from the needles of the Sitka spruce tree, which he was the first European to document and collect seed samples of. Once, when the brewers couldn't find the right kind of spruce, their resident botanist provided the solution:

> They brought me word that there was none of that particular Spruce from which they used to Brew to be found near the landing place, on which I recommended another species (Pinus Canadensis[222]) which answered equally well and made very salubrious and palatable Beer.[223]

He delighted in finding a new species of raspberry.

On a few occasions the crew were afflicted with accidental food poisoning. One might expect this would come from eating incorrectly identified berries or mushrooms but on at least one occasion it happened when a pouch of tobacco fell into the stew. A more serious incident occurred when some crew in a small boat from *Chatham* were poisoned by mussels – one man, John Carter, died. Vancouver named the place he was buried Carter Bay, the place where the mussels were eaten, Poison Cove, and the passage to it, Muscle Canal (now Muscle Inlet).[224]

Not all maladies were digestive, or the result of falling off the rigging, or exploding firearms. Some were the relatively harmless yet aggravating consequences of outdoor adventure – aches, pains, and the occasional infestation of frisky fleas:

> We no sooner got to the Water side than some [of the shore party] immediately stripped them-selves quite naked and immersed their Cloth, others plunged themselves wholly into the Sea in expectation of drowning their adherents, but to little or no purpose, for after being submersed for some time they leaped about as frisky as ever; in short we towed some of the Cloths astern of the Boats, but nothing would clear them of this Vermin till in the evening we steeped them in boiling water.[225]

And not all physical harm was by accident. Some were the result of, in Archibald's words, 'punishment of a very unpleasant nature': Vancouver had Pitt flogged for insubordination again, the boatswain was flogged for conniving with Pitt, the ship's armourer was flogged for drunkenness, and a negligent sailor received three dozen lashes of the cat o' nine tails when *Chatham* ran aground.

One of *Discovery*'s carpenters, last seen opening a porthole while at sea, is presumed to have committed suicide.[226] There was of course little Archibald could do about that.

On another occasion, at Nootka, Archibald was urgently summoned to view the body of a boy from one of the Spanish ships who had been missing for a few days. The body had been found in the woods, a short distance from the inner port of the cove:

He was brought to the Village where I had an opportunity of inspecting his Body, and found that his throat and the right side of his neck had been cut and mangled in a dreadful manner; there were some deep gashes on the inside of his thighs, and apparently a small piece cut out of the Calf of each Leg ... [227]

The cause of death, he concluded, was murder. But being only the coroner, and not a detective, the case was never solved.

Guns and arrows

The closest Archibald ever got to losing any patients occurred in August of 1793 on the shores of what today is southern Alaska, about 180 km north of Prince Rupert. A routine survey of the coast by small boat was under way, one of hundreds they had performed over the previous two summers. Some local Tlingit people arrived on the scene, and the small boats began to trade and exchange gifts with them – as they had done before, dozens of times, without incident. Archibald was in one of the small boats from the *Discovery* with some officers and crew. Four canoes containing about 36 people approached his boat,

> ... caroling and holding up Sea Otter Skins with all the alluring signs of friendship; as soon as they joined us they threw two of these skins into the boat and took the first things that were offered in return with apparent satisfaction without driving a bargain for them as the Natives we had hitherto met with generally did.

Vancouver was in another of the *Discovery*'s small boats nearby, with Puget, conducting their navigational survey. More canoes arrived on the scene and the two small boats were soon nearly surrounded. Had he been watching Vancouver's body language, Archibald might have sensed he was beginning to feel crowded and vulnerable.

Anxieties were raised further when one of the Tlingit men reached for one of the muskets in Vancouver's boat. Archibald had observed earlier that there were more firearms in the area than when he had been here before on the *Prince of Wales* and noted that several were inferior to those of English manufacture 'as their Barrels were secured to the Stocks by means of Iron hoops' rather than brass. He suspected, as did Vancouver, that commercial traders in the area were responsible for the escalation in arms trade and had been driving hard bargains in exchange for low quality weapons. Now, they received confirmation:

> This Indian [wrote Vancouver later] by means of signs and words too expressive to be mistaken, gave us clearly to understand, that they had reason to complain of one or more muskets that they had purchased, which burst into pieces on being fired: a fraud which I know has been practised too frequently, not only on this coast, but at the Sandwich [Hawaii], and other islands in the Pacific Ocean. These defects have not arisen from ignorance or mismanagement on the part of the Indians, but from the baseness of the metal and imperfect workmanship of the firearms.

He added, 'we had reason to suspect that they had been ill-treated in their traffic with white men.'[228]

Vancouver had always been reticent to trade in weapons and resisted the plea to do so now. However, clearly seeing that there were quality British muskets in the boats, the Tlingit men continued to press Vancouver to trade for them. He again refused. They then decided to help themselves. Now things spiralled desperately out of control. The natives, feeling they had been wronged and cheated by previous European traders, grabbed for the weapons and whatever else they could take from the boats. If they were not given justice, they would take it.

Vancouver decided it was time to make a hasty departure and ordered both boats to retreat. But 'a young man, appearing to be the chief of the party, seated himself in the bow of the yawl, and put on a mask resembling a wolf's face' and directed his group to grab the boats' oars to prevent them from leaving. The Europeans were now immobilised, and surrounded by 50 angry men with spears, daggers, and some muskets. The Europeans responded by raising their weapons too.

Now, each was pointing a weapon at another.

Vancouver tried to calm the situation and achieved some brief success. But before temperatures could properly cool, his diplomatic efforts were confounded by the vociferous efforts of a female member of the tribe (in Vancouver's words, 'their female conductress') who 'seemed to put forth all the powers of her turbulent tongue to excite, or rather to compel the men, to act with hostility towards us'.[229]

Responding to the woman's plea, one of the warriors raised his musket, aimed it between Vancouver's eyes, and pulled the trigger … to no effect. Whether it was a lucky misfire, or the man was making a point about the poor quality of the arms he had purchased, Vancouver didn't care to say.

Other warriors then let loose with their spears – one sailed past Archibald's ribs and lodged firmly in the ribs of the wooden boat instead. Another spear flew directly towards Seaman Robert Betton's chest but quick reactions and an instinctive desire for self-preservation allowed him, at the last minute, to deflect the blow sufficiently to save his own life. But then a second spear hit him in the thigh, and he went down. Within seconds Seaman George Bridgeman also caught a spear in his thigh – delivered with such force that the tip passed clean through, from one side to another. In an instant Vancouver's mind flashed back the fateful day at Kealakekua Bay where his hero, Captain Cook, was stabbed to death, and he recalled also the recent loss of *Daedalus's* Captain William Hergest, cut down at Waimea Bay. Fearing they were seconds away from the same fate, Vancouver gave the order to fire:

> Seeing no alternative left for our preservation against numbers so superior, but by making use of the coercive means we had in our power, I gave directions to fire; this instantly taking effect from both boats, was, to my great astonishment, attended with the desired effect, and we had the happiness of finding ourselves immediately relieved from a situation of the most imminent danger.[230]

Now able to grasp their oars, the small boats rowed away from shore as fast as possible, desperate to get beyond the range of the spears and arrows. About a quarter mile from shore they rested so that Archibald could safely attend to the wounded. Betton and Bridgeman were pale and their blood mixed with the salt water sloshing in the bottom of the boats making a grisly scene, but Archibald fashioned tourniquets and bandages to stem the crimson flow. 'I had the satisfaction'

Vancouver wrote later, in his dry understated manner, 'to learn from Mr. Menzies, after he had dressed the wounds, that he considered neither of them likely to be attended with any present danger'.[231] Archibald's actions in the small boats, cool and effective under fire, surely saved the two crewmen's lives. But the incident was a stark reminder of the dangers that accompanied his role, and the unwelcome and regrettable consequences of increased commercial trade in the area.

After reaching the safety of *Discovery* four days later, Archibald discussed the incident with Vancouver and the other officers present as they tried to figure out what had gone wrong. The blame, they felt, lay with the unscrupulous traders who, like the Etches brothers and other commercial speculators, had been incentivised by Cook's reports to seek a fortune in the fur trade. Vancouver summarised their attitude towards the traders in his journal:

> Many of the traders from the civilized world have not only pursued a line of conduct, diametrically opposite to the true principles of justice in their commercial dealings, but have fomented discords, and stirred up contentions, between the different tribes. In order to increase the demand for these destructive engines [the muskets]. They have been likewise eager to instruct the natives in the use of European arms of all descriptions; and have shewn by their own example, that they consider gain as the only object of pursuit; and whither this be acquired by fair and honorable means, or otherwise, so long as the advantage is secured, the manner how it is obtained seems to have been, with too many of them, but a very secondary consideration.[232]

In this regard, Archibald was in complete agreement with Vancouver.

On his charts, Vancouver named the place of their retreat Escape Point, and a nearby island was named for the twice-wounded Betton. Despite his sympathy for the wrongs done by commercial traders who had so aggrieved the natives there, he was nevertheless convinced they had nearly been killed in a premeditated ambush, so he named the place of their encounter Traitor's Cove. As Archibald had already learned, Vancouver was quick to put people into simple categories – friend or foe, loyalist or traitor.

Archibald had met many fascinating people in the Pacific Northwest.[233] Up and down the coast he met Indigenous peoples from many tribes: the Tsimshian, Tlingit, Haida, Eskimo, Tanaina Athabaskan Kwakwaka'wakw, Nuu-chah-nulth, Makah, Coast Salish, Chimakum, Quileute, Chinook, and Tillamook. He had renewed his personal friendship with Chief Maquinna and his sister-in-law, had witnessed the engagement of Maquinna's daughter to Chief Wickaninnish's fourteen-year-old son, had come to know Maquinna's brother Wagh-elas-opulth and a young chief named Nannacoos. At Nootka Sound he met many engaging Spanish officers: Quadra, Galiano, Valdes, Malaspina, and others. In the Strait of Juan de Fuca he met American trader Robert Gray. In Alaskan waters he met Russian traders and dined with them at their Port Etches settlement. There had been many wonders in the West Coast; but after four years in the Pacific Northwest, it was time to head home.

In October of 1794 they headed south towards Cape Horn, Chile. It would take nearly a year to complete the journey home, and it would test his relationship with Vancouver in ways he could never have imagined

Nine

HOME AGAIN, AND AGAIN (1795–1802)

When Archibald finally sailed up the Thames and disembarked from the *Discovery* at Long Reach, Dartford, the ship's sails were faded, frayed, and patched, and her ropes were thin. She was the perfect metaphor for her exhausted crew. The date was 20 October 1795 and they had just completed the longest surveying expedition in history – taking over four and a half years to complete and covering a distance of approximately 105,000 km (121,000 km, if the thousands of miles covered in the many small boat expeditions are included).

Archibald's botanical achievements were massive in both scope and scale. His contributions to the study of natural science were sufficient to fill volumes, and English gardens would be changed forever. From Australia he had sent cones of the Banksia flower, from the Pacific northwest he brought back a species of ornamental gooseberry (*Ribes speciosum*), the Douglas spruce tree (*Pseudotsuga menziesii*), the Nootka cypress tree (*Chamaecyparis nootkatensis*), the Menzies cypress (*Picea sitchensis*), the arbutus tree (*Arbutus menziesii*), and from Chile, the monkey tree (*Araucaria araucana*), to name but a few.

His medical achievements were equally impressive. Taking over the surgeon's role just eight months into the epic voyage, he had not lost a single crewman to illness.

As an adventurer he had seen the beauty of a sun rise over the ocean from the top of a volcano, magnificent waterfalls, whales in the water and eagles in the sky … and he had also seen the ugliness of men shot, flogged, mutilated, and murdered. He had achieved his personal dream of following in the path and style of Cook and, while no one could possibly match Banks, he had now achieved a professional status and reputation that took him beyond that of mere protégé. He was tired, but satisfied that he had proven himself as a physician, a botanist, and a man of action and adventure.

The voyage had delivered many highs over the past few years, but unfortunately it was the lows that were hanging over Archibald and the crew as they eased into port.[234] That morale was low would surprise no-one who had followed the voyage closely. Vancouver's treatment of Pitt was dramatic but unexceptional – by the time the voyage was over, nearly half the crew had tasted the lash at some point or other – and at least two members of the crew had deserted, jumping ship in South Africa and in California. One crew member had killed himself by crawling out of a porthole while at sea. The mental stress of the mission was quite as demanding as more traditional dangers.

Conflict with Vancouver

Yet, despite the intimidating authority Vancouver held over his crew, Archibald never shied away from venting his frustrations directly to Vancouver when necessary. Their conflicts about care of the custom-built plant frame were frequent and legendary. At one point in the voyage, Archibald was so frustrated by Vancouver's belligerence he put pen to paper hoping the written word might have better effect than words which often seemed to have fallen on deaf ears:

Discovery Nov: 18th 1793

Sir,

It is really become so unpleasant to me to represent to you verbally any thing relative to the Plants-frame on the Quarter-deck that I have now adopted this method to mention to you all the alterations or rather additions which I wish to be made to its original plan, for the security of the plants within it, together with the occasional aid that may be required to look after it, in my absence; that my solicitations for its success may not subject me hereafter to such treatment.[235]

Because Archibald's botanical mission was under command of the Secretary of State (through Banks) he had both the right, and the confidence, to press Vancouver for these various accommodations.

But on matters unrelated to botany, he had little say and less clout. He could only vent his opinions and seek remedy privately in correspondence to others. One such occasion was when Broughton sailed back to London on *Daedalus* thus leaving *Chatham* without a commander. Johnstone, assigned to *Chatham* from day one and now a full lieutenant, was the obvious choice to replace him. But Vancouver promoted Puget, of *Discovery*, to the role instead. Archibald wrote to Banks, seeking justice for Johnstone, characterising this event as one of the 'many other strange things in this voyage'. He knew no one else on board could raise the matter with the Admiralty, on punishment of insubordination or worse, so Archibald appealed to Banks to work his magic behind the scenes: 'may I solicit your friendly offices in procuring for Mr. Johnstone the command of the *Chatham*' he wrote, adding 'he has no friends in the Navy except what his merit procured him and being hitherto in a subordinate capacity you may reasonably suppose there are but few'.

Archibald also clashed with Vancouver over his journal. As they entered the final months of their voyage, Vancouver was gathering and preparing his notes so that he could write – as Cook and others had done before him – an authoritative account of their voyage. It would surely be a best-seller. He ordered everyone to hand over their journals and they all dutifully, if unenthusiastically, complied … except Archibald. He refused because, he maintained, his journals belonged to the Secretary of State and not to the Admiralty. His mandate letter from Banks explicitly said his journal was 'to be delivered to H.M. Secretary of State or to such person as he shall appoint to receive them'.

Vancouver was livid. But Archibald held firm. Recording the exchange in a note to Banks, he wrote: 'Thought Captain Vancouver made a formal demand of my journals, etc., before he left the ship; I did not think myself authorized to deliver them … til I should hear from you or the secretary of State for the Home Department'.[236]

For nearly five years Archibald and Vancouver had managed to keep their occasional, and unavoidable, conflicts civil and private. But a few months before arriving home, as they were making their way from St Helena in the South Atlantic, they had an epic, public, row over the plant frame. In the middle of a storm, Vancouver had re-assigned one of Archibald's botanical helpers to other duties. Consequently, the plant frame – long a source of friction between them – was left unattended and many of the plants Archibald had tirelessly acquired and nurtured for the king's garden at Kew were destroyed. Now it was Archibald who was livid. He complained to Vancouver who 'immediately flew into a rage and his passionate behaviour and abusive language on the occasion prevented any further explanation – and I was put under arrest because I would not retreat my expression while my grievance remained unaddressed'.

Archibald took the view that he was acting within his rights, as he had sole authority over matters relating to botany, and knowing that Vancouver had been ordered by the Admiralty to support him in that mission. But Vancouver took the view *he* was acting within *his* rights, being in sole command of the ship's personnel … including its surgeon, Archibald.

They were at loggerheads, but Vancouver had the upper hand, and played it. Archibald broke the news to Banks in an urgent letter: 'I beg leave to inform you that I am now under arrest since the 28th of July last for insolence and contempt as it is termed and as Captain Vancouver will no doubt report me as such on his arrival at the Admiralty'.[237] Banks once again put his social network into play behind the scenes and discreetly arranged for the matter to be quietly resolved. He reached out to the Secretary of the Admiralty, Evan Nepean, and suggested he persuade Vancouver to drop the matter and, at the same time, Banks suggested to Archibald that a formal apology to Vancouver might fall on fertile ground. All the characters played their part as scripted by Banks and, four days after the *Discovery* docked in London, the matter was quietly resolved. Vancouver dropped the charges, and Archibald was at liberty to leave the ship and carry on as if nothing had happened.

Anticlimax

Such was the state of affairs when the voyage reached its conclusion. Everyone was tired, and their nerves were frazzled. The voyage ended more in gloom than glory.

If Archibald and the crew envisaged a peaceful, safe harbour when they reached home, here they were also disappointed. As he had foreseen years before ('I thus leaving our country at a moment when towering aspect of public affairs throughout Europe seems to indicate a general War'), war with France had begun shortly after they left London in 1791. It was still going on.[238]

The French monarch who supported the Americans in their revolt against monarchy 20 years earlier had been beheaded in July of 1793. Thankfully the guillotine, having efficiently removed thousands of French nobles' and revolutionaries' heads from their bodies while Archibald was at sea, remained in Paris and had not made its way across The Channel to London.[239] Yet, elements of British society were also feeling and expressing some of the same pains and tensions that had driven the French peasantry, with the help of some radical intellectuals, to violence and revolution.

Nine days after Archibald returned to London a group of activists ambushed George III's carriage near the Parliament buildings, breaking one of its windows and showering the monarch with broken glass. They were angry about the widespread hunger and poverty in the country, caused in part by the expensive war with revolutionary France, and they wanted the king and his government to implement a programme of political reform and bring an end to the war.

Despite these harsh, violent realities, after five years living among remote tribes in small coastal villages, Archibald began to acclimatise himself to the reality that he was now back at the centre of political, philosophical, and other major events. London itself was now home to nearly a million people, with more arriving every month. Within a few years the London Bridge that had served the city effectively since 1209 would no longer be up to the task, and all agreed it must be replaced by a new one. The cost of living had risen since he left. When buying hair powder, he discovered a new tax had been levied even on that! The assault on George III's carriage, the ideas of Thomas Paine and Edmund Burke, and the state of the war with France, were hot topics back at the Marlborough Coffee House. Archibald's social hours alternated between getting up to speed on the unfolding domestic issues of the day, and regaling audiences with his own tales of past adventure in lands which now seemed increasingly far away.

Virology experiments

Fortunately, Archibald did not have to wait long to find a new outlet for his talents.

Just a month after disembarking from *Discovery* he was contacted by the prominent and well-connected doctor James Carmichael Smyth. Smyth was fifty-four years old, had been born not far from Aberfeldy, had studied medicine at the University of Edinburgh, and had just been tasked by the Admiralty, with George III's encouragement, to test his theory that nitrous acid gas could be effectively used to counteract the deadly viral epidemics which thrived on crowded ships. A hospital ship, the *Union*, had been identified as the site for his experiment. But Smyth had no personal experience at sea, didn't know his way around navy ships, and needed an associate to help manage the experiment. When he heard of Archibald through mutual friends, he immediately knew he would be the perfect man for the job.

The deadly virus in question was typhus. Commonly known at the time as 'jail fever', 'hospital fever', and 'ship fever', it was a constant, troublesome, and mortal threat to the navy. It was a major problem for the army, too, since they depended on the navy to transport soldiers to their various destinations. Crowded together on ships, the ranks of Britain's soldiers and sailors were routinely thinned before they even got to their destinations by this deadly fever. Transmission happened when the faeces of infected lice – on their bodies, on their clothes, or in their bed linen – was passed between them. Unfortunately, this detail was unknown to medical science at the time. Smyth thought – and Archibald concurred – that the virus might be transmitted by exhalation of air from the lungs, or perspiration of liquids from the body, and carried in the air. Therefore, Smyth hypothesised, the solution lay in purifying the air. Smyth's experiment would use vaporized nitrous acid to fumigate the ships from top to tail – including patients, their clothes, and the bed linen.[240]

By supervising the experiment Archibald would expose himself to well over 100 persons in close quarters, but he was keen to proceed. He had come to accept that hands-on field work – even such as might expose one to physical danger – was often necessary to make significant discoveries and advances in science and knowledge, for the benefit of society. Besides, after fully reviewing Smyth's theories and consulting with him on the plan, he was confident the experiment would work. Smyth would later write to the First Lord of the Admiralty George Spencer, the 2nd Earl Spencer, [241]

Mr. Menzies, late surgeon to his Majesty's sloop the *Discovery* … very obligingly undertook the management of the experiment … and it is but doing him justice to say that I could not have found a gentleman better qualified, in every respect, for executing so important a trust.[242]

After being fully briefed on all aspects of the experiment, Archibald left London on the 24 November 1795 for the naval base at Sheerness, just 60 km east of London. The morning of the 25th, he met with Admiral Charles Buckner, the commanding officer of the port, and discussed the project. The distinguished sixty-year-old admiral had once been commander of the 64-gun HMS *Prothee* during the Battle of the Saintes and fully supported Archibald and the project with, in Archibald's words, 'politeness and zeal'. Buckner escorted Archibald to the *Union*, formerly a 90-gun battleship but now a hospital ship packed with 200 people 'of which about one hundred and fifty were in different stages of malignant fever, extremely contagious, as appeared evident from its rapid progress, and fatal effects, amongst the attendants on the sick, and the ship's company'. In the previous three months, eight nurses and two washerwomen had been attacked with fever and three had died.

The next day Archibald returned to the ship ready to begin the experiment. His extensive notes describe his actions:

On the forenoon of the twenty-sixth I went again on board the *Union*. I first ordered all the ports and scuttles to be close shut up; the sand, which had been previously heated in iron pots, was then scooped out into the pipkins[243] by means of an iron ladle, and in this heated sand, in each pipkin, a small tea cup was immersed, containing about half an ounce of concentrated vitriolic acid, to which, after it had acquired a proper degree of heat an equal quantity of pure nitre in powder was gradually added, and the mixture stirred with a glass spatula, until the vapour arose from it in considerable quantity. The pipkins were then carried through the wards, by the nurses and convalescents, who kept walking about with them in their hands, occasionally putting them under the cradles of the sick, and in every corner where any foul air was suspected to lodge. In this manner we continued fumigating, until the whole space between decks was, fore and aft, filled with the vapour, which appeared like a thick haze.

I however proceeded in this first trial slowly and cautiously, following with my eyes the pipkins in every direction, to watch the effect of the vapour on the sick, and observed that at first it excited a good deal of coughing, but which gradually ceased; in proportion as it became more generally diffused through the wards; this effect appeared indeed to be chiefly occasioned by the ignorance or inattention of those who carried the pipkins, in putting them sometimes too near to the faces of the sick by which means they suddenly inhaled the strong vapour, as it immediately issued from the cups.

In compliance with Doctor Smyth's request, the body-clothes and bedclothes of the sick were, as much as possible, exposed to the nitrous vapour during the fumigation and all the dirty linen removed from them was immediately immersed in a tub of cold water, afterwards carried on deck, rinsed out, and hung up till nearly dry, and then fumigated before it was taken to the wash-house: a precaution extremely necessary in every infectious disorder. Due attention was also paid to cleanliness and ventilation.

At the end of the process 'the vapour having entirely subsided, the ports and scuttles were thrown open, for the admission of fresh air. I then walked through the wards, and plainly perceived that the air of the hospital was greatly sweetened, even by this first fumigation'.[244]

Archibald was pleased to report that after two days of this procedure, ' … a pleasing gleam of hope seemed now to call: its cheering influence, over that general despondency which was before evidently pictured in every countenance, from the dread and horror each individual naturally entertained of being, perhaps, the next victim to the malignant powers of a virulent contagion'.[245]

Archibald's belief in the value of hands-on fieldwork, and his having been trained by Hope to use all his senses – not just his eyes but also his ears and nose – paid off further as he walked among the ship's decks. The fumigation had cleansed most of the air, but he detected an awful stench, still, from the on board toilet – which, using the language of the day, Archibald referred to in his notes as 'the necessaries'. Such olfactory observation could only be made by someone on the scene, and because of it he added another dimension to Smyth's experiment: he asked that the toilets be moved outside, writing:

> But there was, in particular places, a constant source of bad smell, which was not easily overcome, and which was occasioned by the necessaries. These were badly constructed. being placed within the ship, to the number of seven on the lower deck, and two on the middle deck, with small funnels that pierced the sides of the ship in a slanting direction, and generally retained the soil[246], unless where a person constantly attended to wash it away, a very troublesome and dangerous office, which chiefly fell to the lot of the nurses, and doubtless tended to spread the contagion amongst them.[247]

He mentioned this to the commanding officer, and three weeks later was able to report,

> My proposal was to remove all the necessaries from the inside, and have them rebuilt on the outside of the ship, and by cutting down the lower edge of the same number of port-holes, to form entrances into them from the hospital, by which they would be equally easy of access to the sick, and the nuisance would be totally removed. This I was happy to find the carpenters were now executing, and I am confident it will be attended with beneficial effects, by rendering the hospital much sweeter, and consequently more agreeable and healthy, both to the sick and attendants.[248]

The experiment was going well enough, and the problem so serious, that a visiting Russian admiral asked to have it repeated on several of his ships. Archibald wrote that

> Soon after my arrival at Sheerness, I had the honor of being introduced to his Excellency Admiral Hannikow, Commander of the Russian squadron at that port, on which occasion he was pleased to express a particular desire of having the most sickly ships of

his squadron purified, by the same process of fumigation, as I was then carrying on, on board the *Union* hospital ship. This being made known to the Lords Commissioners of the Admiralty, they were pleased to declare their approbation … [249]

So, in addition to his work on *Union*, Archibald also oversaw the repetition of the experiment on the 74-gun Russian ship *Pamet Eustaphia*, the 66-gun *Ratvezan*, the 66-gun *Pimen* (her crew so sick other ships were forbidden from coming alongside her, from dread of infection), and the *Revel*.[250]

Putting his nose to good use again, Archibald also advised the Russians to get rid of their smelly sheepskin coats. The coats were well-suited to the dry cold of Russia but were ill-suited to the wet British weather, retained moisture, and gave off a bad smell – a sure sign of impending ill-health. Rather than take insult, the Russians eagerly complied with all of Archibald's recommendations and, subsequently, the senior officer of the Russian fleet, Vasily Chicagov,[251] would write, on 10 March 1796, 'It has been observed that the fumigation, with the nitrous acid, introduced by Mr. Menzies on board the ship *Pamet Eustaphia*, has produced, in a short time, the best effect in stopping the progress of the fever and other evils which were then evidently increasing'.[252]

Around this time, Archibald was also pleased to reacquaint himself with Dr Gilbert Blane, the former physician to the fleet aboard Lord Rodney's flagship during the Battle of the Saintes. Blane was now a Commissioner on the Sick and Wounded Board of the Admiralty and was advising the government on the development of what would become the Quarantine Act of 1799. The two likely enjoyed reminiscing about their days on *Formidable* and catching up on the latest in naval medicine – particularly their work on treatment of infectious diseases. Blane, with his colleague Dr James Johnston, had been recommending sailors' clothes be fumigated with brimstone (sulphur). He and Johnston would later publish their thoughts on this topic in a document called *Letters, Etc., on the Subject of Quarantine*, and would include reference to the fumigation technique espoused by Smyth and executed by Archibald as a possible method for treating yellow fever.[253]

Mingling in the medical community of the day, Archibald would have been fascinated by Edward Jenner's ground-breaking work fighting the scourge of smallpox. He undoubtedly knew Jenner through his network of natural scientists, had probably met him a few times socially, and certainly would have been familiar with the work Jenner had done to classify botanical specimens gathered on Cook's first voyage. Jenner had been invited to join Cook's second mission but was involved in too many other pursuits – studying human blood, the habits of cuckoo birds, geology, development of a hydrogen balloon – to limit himself to the confines of a ship and the limited field of botany.

Smallpox, a scourge as deadly – or deadlier – than typhus was responsible for many thousands of deaths among the general population every year. As many as 20% of the deaths in urban centres were caused by smallpox. Those who survived, anywhere from 20–60% of those afflicted, were left with disfiguring scar-tissue, and a third struck were blind. Like typhus, smallpox spread both virally and through contact. But Jenner's approach was to inoculate the body by vaccination, not cleanse the air by fumigation, and his deliberate use of vaccination to control an infectious disease applied a degree of scientific legitimacy to the procedure that had heretofore been lacking. He performed his first vaccination just five months after Archibald finished Smyth's fumigation experiment.[254]

Archibald likely found Jenner's process intriguing and far advanced over the crude process – variolation – he had used to inoculate Tooworero years before.

He may have discussed Jenner's experiment with another colleague in the medical profession, Everard Home. Home, thirty-eight years old, and just two years younger than Archibald, was assistant surgeon at the Naval Hospital in Plymouth and had been a Fellow of the Royal Society for the previous ten years. Home was also encouraged by Jenner's experiment but worried the sample size was too small: 'If twenty or thirty children were inoculated for the Cow pox and afterwards for the Smallpox without taking it, I might be led to change my opinion'.[255]

Setting their conversation about viruses aside, Archibald and Home spent most of their time focussed on a paper they were writing for the Royal Society. Entitled 'A Description of the Anatomy of the Sea Otter', the paper discussed and described in precise detail the physical attributes and characteristics of the otter using two samples obtained by Archibald near the Queen Charlotte Islands (Haida Gwaii) during his voyage with Vancouver. They delivered their findings to members of the Royal Society, to great appreciation and fascination by the prestigious and accomplished members, on 26 May, just two weeks after Jenner's vaccine experiment.

Pitt's revenge

Meanwhile, if Archibald's eyebrows had been raised by the attack on the king's carriage a year before, he would have been completely bowled over to hear that former *Discovery* midshipman Thomas Pitt had physically attacked their former commander in the street! It seems kicking Pitt off *Discovery* early had been a big mistake. It had given the angry young midshipman time to get back to England well ahead of Vancouver and – more importantly – it allowed the twenty-year-old Pitt to discover that he had inherited the title of Right Honourable Lord Camelford, Baron of Boconnoc. Vancouver had unwittingly thrashed and humiliated one of the wealthiest and most powerful men in England. Even worse, Pitt had never managed to get over the humiliation of his flogging. Instead, he nursed his grievance and his thirst for revenge, becoming angrier by the day.

The moment Pitt heard the *Discovery* was in back in London, he sent Vancouver a letter challenging him to a duel and promising that, should he not accept the challenge, he would see to it that 'the few surviving remnants of your shattered character' would be lost forever. Pitt, showing off his wealth, even enclosed a draft for £200 (a comfortable annual salary for some people) to cover Vancouver's travel costs to attend the duel. How the tables had turned. At sea Vancouver had been the omnipotent one. Now, on land, Pitt was the powerful force. Vancouver wrote back to Pitt refusing the challenge on the grounds that when he had administered punishment he had simply been following the rules – it was nothing personal – and suggested Lord Camelford take the matter up with the Admiralty.

It was a nice try, but the zealously proud Pitt was not so easily assuaged. Having failed to get his duel, he followed Vancouver to his home in Petersham, Surrey, and confronted him face to face with a renewed demand for satisfaction. Vancouver frantically wrote letters to the Foreign Secretary and others, and initiated legal action to keep Pitt at bay. But on Wednesday 21 Sept 1796, Pitt saw Vancouver and his brother Charles walking up London's Conduit Street, on their way to the Lord Chancellor's house at Bedford Square. He pounced. Years of anger raged through his veins and, in full view of the startled public, he levied several blows at Vancouver with his walking stick. Vancouver parried with his own stick while brother Charles first grabbed Pitt by the throat, then delivered several body blows with his fists.

The street melee between the Lord and the commander ended quickly enough, but the event lingered on painfully in the public arena and soon inspired a devastating cartoon by notorious caricaturist James Gillray.[256] Archibald's jaw would have dropped as he looked at the cartoon and saw a trim feisty-looking Pitt with his cane raised and a speech bubble over his head saying,

> Give me Satisfaction, Rascal! - draw your Sword, Coward! what you won't? - why then take that Lubber! - & that! & that! & that! & that! & that!

. . . while a portly and somewhat terrified-looking Vancouver, arm raised in self-defence, replies,

> Murder! - Murder! - Watch! - Constable! - keep him off Brother! - while I run to my Lord-Chancellor for Protection! Murder! Murder! Murder!

It made them both look foolish but as a piece of 18th-century media manipulation, it was sublime: Pitt was fulfilling his promise to shatter Vancouver's reputation.

No doubt Archibald's scientific colleagues and regulars at the coffee house peppered him with questions about the scandalous public spat. It would have been common knowledge that he had spent years in close quarters with both men and had tales to tell. Who knows to what degree he obliged, but he was likely discreet and certainly aware it was far better *not* to be either gentleman's enemy. He may have looked back on his own past confrontations with Vancouver and thanked his lucky stars they had never come to blows in the street or threatened to settle their plant hutch disputes with pistols at dawn.

Rather than gossip about Vancouver and Pitt, Archibald spent his time productively. He had intellectual discussions with James Edward Smith at the Linnean Society, with Banks at his home, and other old and new contacts in the natural and medical sciences. He stayed as far away from politics and power-dramas as he could, preferring instead to visit Mr Aiton at Kew Gardens where he was pleased to see the rumex (*Rumex giganteus*) seeds he had sent from Hawaii were prospering. And he began work editing his journals for publication. He was also pleased to accept the Admiralty's appointment as surgeon to HMS *Princess Augusta*.[257] The role aboard the Thames-based 20-metre-long yacht came with no significant demands on his time, but it increased his pay, kept him on active service, and allowed plenty of liberty for him to continue the important medical pursuits that might benefit the Admiralty.

The undemanding assignment to *Princess Augusta* also bought him some time to consider his future. The government appeared to have no interest in establishing a settlement at Nootka, as he had once thought. Britain was busy fighting a war with France and managing domestic issues, and so long as Spain had abandoned its proto-settlement at Nootka, Britain was satisfied to let sleeping dogs lie. The calculation would have been much different had Vancouver discovered a Northwest Passage, but, his extensive charting of the coast had dis-proven its existence. So, Archibald's dream of returning there came to an inglorious end.

He may have been disappointed that a return to the beauty of the West Coast was not in the cards; however, he was by now probably enjoying his time in London and likely thrilled to be back among the company of many welcoming peers in the medical and natural sciences. Days were filled sharing his experiences with other scientists, cataloguing and discussing his findings, and comparing his discoveries with those of others returning to London with discoveries of their own from across the country and around the world.

Janet Brown

His return to London allowed Archibald to engage in a much different social life than he had experienced aboard the *Discovery* and he was introduced to new and interesting people on a regular basis. A notable new acquaintance was a woman named Janet Brown. The wispy twenty-six-year-old was the older sister of Adam Brown – a veteran of HMS *Chatham* and trusted friend of both Archibald and Johnstone. Archibald was immediately struck by her progressive attitude and contemporary style; she was charmed by his gentle manner and insatiable curiosity about nature and the world.

Janet was a lively and modern woman. She wore her brown hair in short curls and, like many of her contemporaries, she had gleefully abandoned the rich silks and firm conical corsets of the past in favour of the neoclassical dress style with its high waist and lightweight diaphanous flowing cotton. The new fashion reflected the events of the past decade – the French revolution had triggered a new interest in classical style, with dresses that imitated the clothing of ancient democracies. General Napoleon Bonaparte's adventures in Egypt, the battles of the British Royal Navy, tales of exotic adventure in lands far away, were all influencing the fashion, architecture, and furnishings of the day, and Janet herself was no exception.[258] She was a breath of fresh air – bewitching to behold and pleasing to be around. She was, to Archibald, a living exemplar of the age: bold, free, passionate, confident, and bursting with new ideas.

In addition to candle-lit dinner parties with friends, Archibald would have accompanied Janet to various events and activities around London. They may have taken in the comic Opera *The Duenna*, written by the Irish wit Richard Sheridan, with a score composed by English musician Thomas Linley, at the Convent Garden Theatre. The Opera had first been performed twenty years earlier, but it was a popular favourite, one of the most successful operas ever staged in England, destined to be performed again and again. The story of the Opera was set in Spain, and Archibald may have used it as an opportunity to amuse Janet with his own real-life stories of days spent with Juan Francisco de la Bodega y Quadra at Nootka, or the time in Chile when he had been entertained by its colourful Governor Ambrosio Bernardo O'Higgins de Vallenar.

Had they attended the 2 Oct 2 1798 performance of *The Revenge* by Edward Young at Theatre Royal on Drury Lane, they would have witnessed an event similar to the one Rodney's daughter had written about years earlier. At the end of the performance the audience received a dramatic surprise – an excited voice announced that Rear Admiral Horatio Nelson had defeated the French fleet at the Battle of the Nile! The audience immediately erupted in cheers of 'Huzzah for Nelson' – just as they had once cheered Rodney upon receiving news of the Battle of the Saintes. Vivid memories of the naval battle glory, and carnage, would have flooded Archibald's senses as he and Janet joined the throng in loud voice as the orchestra played 'Rule Britannia', and 'God Save The King'.

On another outing – perhaps sharing a picnic hamper stuffed with treats from Fortnum & Mason – they may have shared an animated discussion about the outrageous new 'top hat' introduced by London haberdasher John Hetherington. According to the newspaper reports,

Hetherington's hat, when first seen on the street, was so outrageous that several women fainted, some children screamed, dogs yelped, and an errand boy's arm was broken when he was trampled by the mob.[259]

Time spent with Janet would have been enchanting, and their social engagements would have been full, but these distractions had also pulled Archibald away from his journals. After three years he still had not completed editing them. From his cousin's comfortable quarters at 31 Berkeley Square, Archibald responded to Banks's latest enquiry: ' … thanks for your friendly admonitions and solicitations respecting finalising of my journal before Captain Vancouver's is published – It is what I most ardently wish, for more reasons than one, and have therefore applied to it very closely … '[260] But, to Banks's chagrin, the motivation to publish before Vancouver soon vanished: just four months later Vancouver died, at age forty – not by the hand of Pitt, thankfully, but of natural causes.[261]

Flagship doctor

Vancouver's death was a stark reminder to Archibald that he was not getting any younger. He now determined to change his career and to settle down. The first step in his plan was to transition from naval surgeon to civilian doctor. This was achieved with the help of Dr Blane who, along with his colleague Dr Johnston, recommended Archibald for an MD degree, which was duly awarded to him by King's College, Aberdeen on 24 July 1799.[262] The next step was one final, naval-career-topping, assignment at sea. This was achieved when the Hon Hugh Seymour, member of Parliament for Portsmouth, vice-admiral, and soon to be commander of the Jamaica Station (and also, in the Eastern Caribbean, commander of the Leeward Islands Station), invited Archibald to serve as surgeon aboard his flagship, HMS *Sans Pareil*.[263]

As one of the Lords Commissioners of the Admiralty from 1795–1798, Seymour had undoubtedly followed Archibald's work on the *Union* and would have been impressed by his association with Blane, Banks, and others. Archibald was precisely the calibre and quality of surgeon he would want for his flagship. It was nearly 20 years since Archibald had first sailed the Caribbean waters but he remembered well the reverence and respect he had had for Dr Blane when serving as Rodney's surgeon. He was deeply honoured, and perhaps just a little bit proud, to be returning to the area in a similar capacity.

The 80-gun *Sans Pareil* was both a grand assignment and a grand ship. She had been captured from the French a few years earlier, in 1794, was twice the length of *Discovery* and her crew was seven times larger – just over 700 men. And while the *Discovery* had at least one Lord amongst its midshipmen, the *Sans Pareil* had so many she was sometimes sarcastically referred to as 'The House of Lords'.

The assignment to *Sans Pareil* was prestigious but also not without an element of danger. The war with France, Spain, and Holland continued, and the waters around the English Channel, the Atlantic, and the Caribbean were active theatres of war. Shortly before Archibald arrived at Jamaica Station, the British forces – there were many battleships under Seymour's command – had captured the Dutch colony of Suriname and also confiscated a French vessel, *Hussar*. Shortly after he arrived, they captured a French privateer, *Pensée*, and the schooner *Sapajon* was captured from their base at the Leeward Island Station. A Spanish trade vessel, the *Guakerpin*, was taken shortly afterwards. Archibald's heart was particularly warmed when he heard the Spanish privateer *Esperanza* had been captured near Cuba by the 18-gun HMS *Lark* … commanded by his old friend James Johnstone.

For two years Archibald sailed the Caribbean and proudly performed all the duties of a senior surgeon to the fleet. He did his best to make sure ships and hospitals were well-ventilated, and free from noxious smells; he prescribed diets that included vegetables and fruits, encouraged the production of various teas and beers made from vitamin- and mineral-rich organics, and he mentored many other young surgeons serving with the fleet. The war continued to run its course, and while the Caribbean theatre was always lively with various strategic manoeuvres and the disruption or protection of trade vessels and routes, the glory battles were mostly held elsewhere in Europe and the Mediterranean. The revolt in Ireland seemed to be resolved – Ireland joined with Great Britain (England and Scotland) to create the United Kingdom of Great Britain and Ireland on New Year's Day, 1801. Archibald may have imagined the Irish felt much the same about the Union as many Scots had felt about the Union of 1707 … but he would have hoped their inclusion in the new United Kingdom would bring prosperity and enlightenment to all its partners. He would have also been sorry to hear that Edward Riou, veteran of Cook's third voyage and shipmate of Vancouver, had been cut down by cannonball at the Battle of Copenhagen in April 1801. The rise of Napoleon to political supremacy in France was also an interesting development – the general had become First Consul in 1799. The world was changing, and so was Archibald. After 20 years in the Royal Navy, he was finally ready to turn the page.

In September of 1802, he came home again.

But this time he was not under arrest. And Janet was waiting.

Ten

LONDON LIFE, AND DEATH (1802–1842)

Within a month of his return to London, Archibald married Janet Brown at St George's Church near Hanover Square. The church, built in 1724, was an impressive structure set in the increasingly desirable Mayfair neighbourhood, halfway between Soho Square and Grosvenor Square. The front portico, supported by six Corinthian columns, projected a grand and welcoming presence to the bride, groom, and their closest friends and family, including Archibald's cousin John Walker, Janet's younger brother Adam and her younger sister Ann.[264]

While Archibald was in the Caribbean, Janet had made all preparations for their marriage and domestic life. Her father, a prosperous London coal merchant,[265] was likely happy to see her wed at last: at thirty-two years of age she was no longer a young maid. If Mr Brown had been eager for the day she wed, he was likely also pleased she had waited to find a good match. Archibald was an Edinburgh-trained doctor with well-placed friends; this would have given Mr Brown confidence that he could provide a stable and comfortable life for his beloved daughter.

The newly-weds celebrated their union by commissioning miniature portraits by Thomas Richmond at his studio at 42 Half Moon Street.[266] For the occasion Archibald wore a high-collared blue velvet frock coat, and a white linen shirt with matching neck scarf. In his waistcoat was the new 18 carat gold watch he had purchased days earlier from watchmaker John Grant of Fleet Street – a prestigious timepiece for the newly wed doctor. According to the trend of the day, Richmond's assistant clipped a lock of Archibald's light-brown hair to be woven into a decorative display under the portrait's glass-backed case, then applied a light dusting of powder for the formal portrait. Janet dressed in a simple all-white ensemble, a loosely flowing garment reminiscent of the robes worn by classical philosophers. Her brown hair was cut stylishly short and from her ears dangled ruby earrings, each one like a single large raspberry. They chose a red leather case for her portrait.

The two settled into life at No. 6 Chapel Place in Cavendish Square, just a five-minute walk north of St George's Church. For their first Christmas, Janet would have decorated the walls and ceiling of their dining room with boughs of living green – crisp leaves of holly, mistletoe, and ivy around the wall and a mighty blaze roaring up the chimney. Archibald's days of tropical Christmases in Hawaii and elsewhere were over, but they would certainly be cosier than ever before. Forty-eight-year-old Archibald transitioned from bachelor to husband, and from navy surgeon to gentleman doctor, with surprising ease. His association with Sir Joseph Banks, Admiral Rodney, Dr Blane, Everard Home, Lord Seymour, his reputation for having kept those under his care alive, and his medical research tackling pathogens, all combined to make him a highly qualified and popular physician.

In addition to his renowned medical abilities, patients were also attracted to his colourful personal history. Few could boast that their personal doctor had twice circumnavigated the globe, walked the beach where Cook was slain, battled the French and Spanish in the Caribbean, or climbed a volcano. He was a most desirable doctor, and his practice quickly flourished.

Social life

Archibald and Janet had a lively social life in London, but their best friends were likely the Smiths.

Archibald had long been a friend of James Edward Smith, but as a newly wed London resident the relationship became closer. Janet and Pleasance Smith were of the same age and shared many interests. Pleasance had recently had her portrait painted by artist John Opie, and Janet agreed with their abolitionist friend William Roscoe that 'he who could see and hear Mrs Smith without being enchanted has a heart not worth a farthing'.[267] When the foursome got together Archibald and James would discuss matters of natural science and the affairs of the Linnaean Society – including its move in 1805 to a new location on Gerrard Street, while Janet and Pleasance discussed their interests in literature, nature, art, fashion, and social issues of the day.

The Smiths had moved from the fast-growing city of London to their country home in Norwich, Surrey a few years earlier but were frequently in town. On those occasions the Menzies's were always enthusiastic hosts for the Smiths. A typical letter from Archibald to Smith reads,

> You mentioned in your last [letter] of your intention of coming to town early in April. I can only say that your old apartments are ready for you and that we shall all be happy to see you but you will make yourself still more welcome to us if you bring Mrs Smith with you ... [268]

Clergyman William Fitt Drake, a nineteen-year-old botanical protégé of Smith, reinforced these messages, writing to Smith from 6 Chapel Place to say 'Mr and Mrs Menzies are quite well, and will be exceedingly glad to welcome you ... they enquired very particularly after Mrs Smith ...'[269]

When they could not come in person, the Smiths shared the bounty of their estate with their London friends through frequent letters and parcels. Archibald's letters to Smith are peppered with thanks. The fresh herring Smith had sent was 'the best I ever tasted' the Turkey was 'the finest I ever saw'.

With the help of Janet's social graces and sparkling personality, Archibald blossomed as a generous host and popular raconteur. The Smiths were always welcome as house guests, as were many botanists young and old. From Drake's other correspondence to Smith, it's clear Archibald was still actively also in touch with Banks and, through him and other contacts in society, was well-informed on the events, developments, and adventures of the day. Drake wrote to Smith,

> The first piece of news that I heard from Mr Menzies was that Sir Jos. Banks is the Knight chosen to attend the Duke of Sussex at the ensuing installation of the Knights of the Garter; and by his attendance on a Royal Duke, he is, by royal mandate, to precede every other Knight of the Bath: I dare say this is highly gratifying to Mrs. Banks!![270]

Archibald's medical practice prospered sufficiently to fill his days. At night, Janet's hospitality and Archibald's gregarious charm made their home a place where established intellectuals and ambitious neophytes were keen to be received. The older gentlemen would review academic papers, discuss the correct phylum, genus, and species of plants, and conduct the business matters and membership roster of the Linnaean Society. The younger ones would come for stories of adventure and to learn directly from the botanist of *Discovery*.

Guests and correspondents included Matthew Martin (the English merchant, naturalist, and philosopher); James Dickson (one of the original members of the Linnaean Society and owner of a Herb Shop in Covent Garden); William Townsend Aiton – son of William Aiton, the first Director of Kew Gardens (designer of the gardens at Buckingham Palace); Christian Friedrich Schwägrichen (Leipzig botanist and professor); William Jackson Hooker (professor, botanist, illustrator, and Kew Gardens Director); John Smith (botanist and first curator at Kew Gardens); W. H. Harvey (Keeper of the Herbarium at Trinity College, Dublin); and American botanist Asa Gray. The younger set included botanists for the Hudson Bay Company David Douglas and Meredith Gairdner. Janet's brother Adam and sister Ann were also frequent guests. Adam would eventually get his own command, HMS *Sapphire*, and Archibald – his old mentor and comrade from HMS *Discovery* – would have helped him prepare for his posting to the Jamaica Station.

In 1804 Archibald was again likely startled, but at this point no longer surprised, to hear that Thomas Pitt had finally got himself killed. The event occurred just 4 km west from Archibald's home, at a meadow near Holland House in Kensington. True to form, Pitt had contrived to become offended by someone who wished him no offence then dramatically raised the stakes by demanding satisfaction.

London journalist Pierce Egan reported that 'The memorable duel which deprived Lord Camelford of his life, like the Siege of Troy, was, on account of a woman'.[271] The love triangle was between Pitt (Lord Camelford), Eliza Symons, and Pitt's close friend Captain Best. Symons repeated to Pitt something she claimed Best had said to her. Pitt was offended and demanded an apology from Best. Best, however, denied saying any such thing, and demanded Pitt then apologise to *him*. Then, according to Egan,

> [Pitt's] towering spirit said that it might look like FEAR if he apologised to Capt. Best – also, a sort of 'begging of his life'; and to any other man, it should seem, he would have *apologised*. Therefore, in order that no *imputation* might be levied at his courage, he not only provoked but hurried on the duel, accompanied with words that no gentleman could put up with.[272]

On a crisp March morning the two friends met at Gloucester Coffee House on Oxford Street and made their way from there to a meadow at the Horse and Groom in Kensington. Standing 30 paces apart, Pitt fired the first shot and missed. Best responded, sending a musket-round through Pitt's ribs and lung. Surgeon Simon Nicolson, whom Archibald knew as an assistant to his friend Everard Home, examined Pitt but could do nothing to save him.

As Smith's young clergyman friend Drake might well have observed 'Pride goeth before destruction, and an haughty spirit before a fall'.[273]

A changing world

Although Archibald had married and was enjoying a comfortable professional and social life in London, he was briefly tempted to go back to sea again. In 1805 the Admiralty offered him an appointment to HMS *Zealous*, a 74-gun battleship recently repaired at Portsmouth. The ship was part of Nelson's fleet and was destined for the Mediterranean, where he was hunting the elusive French fleet, which had recovered from the savage beating he had given them seven years earlier at the Battle of the Nile. Archibald turned the offer down, committed to his new role as husband and civilian. But he may have experienced some regret when hearing later of Nelson's victory at the Battle of Trafalgar. News arrived at the Admiralty in November that year, with dispatches from Vice Admiral Collingwood reporting that 'the enemy's ships were fought with a gallantry highly honourable to their officers, but the attack on them was irresistible, and it pleased the Almighty Disposer of all Events, to grant his Majesty's arms a complete and glorious victory'.[274] The action, which employed a similar strategy to that used by Rodney at Saintes, cost Nelson his life, but the victory was so absolute that his name would live on for generations.

Hearing the stories of the epic sea battle, Archibald would have recalled the thrill of survival and would have been pleased the 33 British ships lost only 400 men. But the French and Spanish loss of over 4,000 lives was astonishing, and he could only have imagined the despair of their surgeons.

There were other stories from around the world that made their way back to London. Stories of adventure and discovery, of conflict and conquest. Those Archibald followed most closely had to do with his time spent in Hawaii and along the northwest coast of the American continent.

Archibald would have been very interested in Alexander Mackenzie's account of his voyage to the West Coast over land, *Voyages from Montreal to the Frozen and Pacific Oceans*, when it was published. He was likely amused to learn from Mackenzie's account that after arriving in Bella Coola, he had been informed by the local Indigenous people there, the Nuxalkmc, that two other Europeans – 'Macubah' [Vancouver] and 'Bensins' [Menzies] had recently visited them by boat.

American President Thomas Jefferson also read the account and presented a copy of Mackenzie's book to Meriwether Lewis before he and William Clark began to plan their land route from the eastern side of North America to the Columbia River and the distant western coast, which the *Discovery* and *Chatham* had charted ten years earlier.

Other news arriving from the West Coast was more concerning. Accounts arrived in London of a devastating attack by Chief Maquinna against an American fur trading ship, the *Boston*. The reports said the action at Nootka had 'principally arisen from the imprudent conduct of some of the captains and crews of the ships employed in this [fur] trade, in exasperating them by insulting, plundering, and even killing them on slight grounds'.[275] In other words, they pushed Maquinna too far, and Maquinna pushed back. James Jewitt, a survivor of the attack, wrote that he had been knocked unconscious and fallen below deck during the attack but afterwards Maquinna had led him to the quarterdeck 'where the most horrid sight presented itself that ever my eye witnessed – the head of our unfortunate captain and his crew, to the number twenty-five, were all arranged in a line'.[276]

When Archibald read of this 'Massacre of the *Boston*' at Nootka, he was surely alarmed and disappointed by the violence. But he perhaps was not surprised. He had noted years earlier, with regret, the problems caused by unprincipled and unregulated fur trade in the area, and particularly the problems caused by the trade in weapons. Archibald retained his faith in Adam Smith's economic theories, and the many benefits of international trade, but he had also seen enough of men to know that not all of them traded fairly or ethically. Had the British not been distracted by other events, had they established an official settlement in Nootka, as he once hoped, perhaps some sense of order and accountability would have taken hold to regulate trade and enforce the law.

When Jewitt's full account was published later, Archibald would have read of his experience living in Nootka. He may have enjoyed reading that the remnants of Quadra's house and garden were still in evidence at Friendly Cove, and how Jewitt had also lived in Tahsis. Perhaps Jewitt's account brought back memories of happier days when he had dined with Spanish officers in their fine quarters and where he, Vancouver, and others had dined in such hope and expectation with Maquinna.

More North American news broke in 1812 when, despite having achieved a peace with Great Britain 30 years before, the United States declared war on Britain's eastern colonies in British North America. It was a poor decision, and the war did not go well for the United States. But Archibald would have been proud to learn that during the war one of *Discovery's* former midshipmen, Robert Barrie, had been put in command of the seventy-four-gun HMS *Dragon*, participated at the blockade of Chesapeake Bay, and served as the commodore of the squadron for several months, capturing over 85 vessels. In addition to Barrie's other battle achievements, Archibald would have been most pleased to read that he had been able to liberate ten slaves from tobacco farmer Thomas Whittington. The *Dragon* was anchored off St. George's Island, Maryland, just 6 km off the Virginia shore and, seeing the British flag, the ten Virginian slaves grabbed a few rowing boats and escaped to freedom. Undoubtedly, Archibald applauded when he learned Barrie had been appointed to the Order of the Bath and, eventually, promoted to the rank of rear admiral.

Many of Archibald's old shipmates received high honours, promotions, and titles. Navy surgeons were somewhat of a different breed and weren't eligible to advance in the same way. Nevertheless, as a renowned civilian doctor, with an accomplished navy background, Archibald was honoured with an appointment as apothecary to the Royal Household in 1819.[277] The appointment came with an annual salary, and an obligation to provide general medicalpractice services to the royal household at Buckingham Palace, if called upon to do so. This put him once again in the company of his old friend Blane, who had been appointed physician to the Prince Regent (George III's son, the future George IV) and who had been made a knight baronet seven years previously.

By the 1820's Archibald's professional reputation, both as a physician and botanist, extended well beyond London and Great Britain. In 1822 he was awarded an honorary MD from the University of Leipzig and appointed a member of Leipzig's Natural Research Society.

He received visitors from continental Europe with some regularity, such as the distinguished Austrian physician and botanist Joseph August Schultes who, after his visit, reported that seventy-year-old Archibald was 'now as active as a person of forty, and is in great practise as a surgeon'.[278]

But the most startling and unexpected visitors to London were the King and Queen of Hawaii.

Kamehameha II

The visiting king was Kamehameha II, son of his old friend Kamehameha I. With him was his wife, Queen Kamamalu. The young Hawaiian monarch was just twenty-two years old when he took over from Archibald's friend King Kamehameha I five years earlier.

Like his father, Kamehameha II was keen to build diplomatic relations with the British monarch. So much so that he decided to make a personal visit so they could talk king-to-king with George IV. Archibald would have been as surprised as anyone else when Kamehameha II arrived in London and took well-appointed rooms at Osborn's Caledonia Hotel in the Adelphi district. He had likely never imagined this scenario. But he would have been amused that the whole city now seemed in awe of the Hawaiian king, queen, and their travelling court.

By this time London was home to over 1.5 million people, and with its growth had also become increasingly cosmopolitan and full of people from far away. Londoners now shared their city not just with people from continental Europe, but also from Africa, the Caribbean, India, and a colourful variety of returned colonists, settlers, adventurers, explorers, and traders from North America, Australia, and around the world. It was probably the most diverse and varied urban population in the world. Nevertheless, the Hawaiian royals fascinated everyone.

Crowds gathered outside Osborn's just for a chance of seeing them in person. People were curious about all aspects of the Pacific potentate and wanted to see what he and his queen wore, what they ate, how they talked and behaved. The royal couple posed for portraits by a young artist – and future painter-in-ordinary to Queen Victoria – John Hayter, who quickly issued prints for sale to the insatiable public. There was even a rumour they had brought Captain Cook's bones with them, for presentation to George IV.

The interest in Hawaii was never higher and those who knew of Archibald's adventures there decades earlier peppered him once again with requests for stories of Kamehameha I, his inspection of Cook's bloody shirt, his ascent to the peak of Mauna Loa, the feather cloak he had been denied, and a dozen other stories of days gone by.

The Hawaiian King and Queen had a full social calendar while waiting for their audience with George IV, including attendance at a grand party thrown for them by Foreign Minister George Canning at his Kensington mansion, and even attended a public performance at the Drury Lane Theatre. George IV had set aside his seats for their use, and the packed house was as thrilled to watch the stage performance as they were to watch the reaction of Hawaiians in the Royal Box. The weekly publication *Bell's Life* in London reported on their visit:

> The Royal party from the Sandwich Isles continue to view the curiosities of the Metropolis, not with standing our varying atmosphere, contrasted with their own salubrious unchanging climate, renders their seasoning very severely felt, particularly by the Ladies. Certain females of rank, who were at Drury Lane Theatre on the 4th instant, declared themselves quite fascinated with the manly appearance of King Rhio-Rhio[279] (the travelling name of Tamehemeo,[280] would now, we are told, be deemed uncourteous), and thought him likely to become rival to our Royal Sovereign – possibly his age may be considered an advantage … His Royal Consort had discarded her native plumes for some other ornaments, which became her very well. The acting of Liston called forth in them … external marks of pleasure; else, they conducted themselves with an easy *nonchalance* that astonished John Bull, who is accustomed to gape and wonder at everything … [281]

Unfortunately, the couple also paid a ceremonial visit to the Royal Military Asylum – a school for children of military widows – a place well known for outbreaks of contagious diseases. Within three weeks their entire party was sick with measles.[282] By 14 July, they were both dead. Archibald was most certainly appalled. London was dangerous enough, with its crowded, poorly sanitised, mostly unvaccinated population. But taking the Hawaiians to a school full of children was a ridiculous decision and he surely regretted that the progress made on smallpox vaccination had not advanced to other potentially fatal areas of contagion.

It is unfortunate that Archibald had not also been appointed physician to the Hawaiian Royal Household. He might have been able to persuade King Kamehameha II and Queen Kamamalu not to travel to London in 1824, or at least to avoid visiting certain places there. London was not the healthiest place to visit. In 1832, a cholera outbreak would take 4,000 Londoners to the grave, and more would follow in subsequent years.

Last man standing

Two years after Kamehameha II's visit, Archibald fully retired from medical practice.

Like many in his situation he moved closer to the countryside … though in truth that was not very far: the home he and Janet moved to was only 4 km east of their Cavendish Square home. But it was a new development and much more peaceful and pastoral. The place they chose was Ladbroke Terrace in Notting Hill.

The area may have reminded Archibald of the New Town in Edinburgh 50 years earlier. Until very recently it had been part of a 69 ha estate sparsely populated by tenanted farms. The new development was just west of Hyde Park, slightly north of Holland Park. Estate owner James Weller Ladbroke had observed the rapid population growth in London and began developing his property in 1821. Thomas Allason was engaged to design the development featuring a north-south street bisecting the estate and a series of crescents and gardens on either side. The streets were filled mostly with houses in the 'Georgian terrace' style, and some semi-detached villas, all with carefully designed vistas and 16 communal gardens accessible only to the residents of houses backing onto them.

Archibald and Janet's home at No. 2 Ladbroke Terrace had white Doric order Greek columns framing the front door and a brass knocker not nearly so large as the haunted one famously described by Charles Dickens in *A Christmas Carol* but certainly just as effective. Like the other stylish homes on the street, their new home was 7 metres wide and 10 deep, had two floors and a basement. On the ground floor the entrance hall gave way to a staircase, off which were the dining room, a smaller rear drawing room, and a small study where Archibald was able to manage his correspondence. Upstairs there were four bedrooms and a water closet. All rooms except one bedroom had a fireplace which, like the kitchen range, was fuelled by coal from the brick-paved coal cellar. The basement had kitchens equipped with a stone sink, a dresser and plate-rack, a closet-cupboard, and gave access to a covered area off which were a small pantry, a cistern, and a coal cellar. There was a small area at the rear from which access was gained to a privy and to the steps leading up to the back garden. At the side were two more brick-paved cellars: one for wine, the other for beer. Candles and oil lamps provided lighting after the sun went down.

When Archibald retired to this Ladbroke home he was seventy-two years old, had spent 20 years in the navy and 20 more as a London doctor. He had also outlived many of his contemporaries.

His friend and lifelong colleague Jonas Dryander had died at the Linnaean Society's house in Soho Square in 1810. Captain William Bligh of *Bounty* fame became a vice admiral and dropped dead in Bond Street, a short walk from Archibald's medical practice, on his way to visit his surgeon in 1817. Joseph Baker, the first member of the *Discovery* to sight Mount Baker and Archibald's companion on his historic trek to the top of Mauna Loa died the same year. King George III, the inquisitive monarch whose interests filled Kew Gardens, the Natural History Museum, and the British Library with plants, artefacts, and books, died in January of 1820. Five months later, Joseph Banks, without whom Archibald might never have seen the Pacific, died at his country home. William Broughton, the first European to sail up the Columbia River, was appointed to the Order of the Bath and died in Florence, Italy the next year. Peter Puget, commander of *Chatham* after Broughton, died in 1822, by then both a rear admiral and member of the Order of the Bath.

The greatest loss prior to his retirement was that of James Johnstone. The two had known each other since early days on the *Assistance* and *Formidable* and had shared many hours and days exploring the Pacific Northwest coast in small boats from *Chatham*. They had maintained their friendship and a lively correspondence over the years, and they had each followed the other's careers and lives closely. After capturing *Esperanza* near Cuba, Johnstone had continued to capture Spanish ships and eventually became Commissioner of the Navy in Bombay before retiring to Paris in 1817.[283] Sensing the end of his life was near in 1823, the sixty-four-year-old Johnstone travelled to Archibald's home and stayed under his care and friendship until he passed away there, on April 1.[284] Archibald was with him to the very end of his voyage.

Archibald had no doubt been saddened to see lifelong friends pass away before he retired, but several others who had featured prominently in his life would also precede him: James Edward Smith, a close friend for 40 years, would die just two years later, in March of 1828. Janet's brother Adam Brown, a lifelong friend to Archibald and Johnstone, died the following month.[285] King George IV, who had been Prince Regent during George III's challenging bout with mental illness, passed in 1830. Everard Home, with whom Archibald had collaborated to present an academic discourse on the sea otter, went on to be appointed sergeant surgeon to the king, was made a baronet, and died in 1832. The next year another veteran of *Discovery* would pass: Joseph Whidbey. In the years following the *Discovery* expedition he had gained some fame as an excellent engineer. Former master's mate on *Discovery*, Thomas Manby, advanced to command several battleships in combat against the French, was suspected of sleeping with the Princess of Wales, was promoted to rear admiral and would die from an opium overdose in 1834. Archibald's mentor and friend Gilbert Blane would die in 1834 and be forever known as the father of naval medical science.

But more were to come. Archibald and Janet never had any children, but had they done so, they would likely have expected him to be very much like David Douglas. When Archibald first met Douglas, in 1824, the twenty-five-year-old Scot probably reminded him of his younger self: he was curious, intelligent, inspired by the natural world, and thirsty for adventure. The meeting was so that Archibald could prepare him for an upcoming voyage up the Columbia River. The Hudson's Bay Company had begun to use the Columbia as their gateway to Pacific trade – the river was an important trade route, and its active use by the company helped, along with Broughton's voyage in boats from the *Chatham* 30 years previously, to retain British territorial influence along the West Coast. Douglas's work for them would be important to science, trade, and geo-politics, and Archibald had much advice to pass along.

The two had kept in touch over the years since their initial meeting and Archibald was likely thrilled to follow Douglas's progress. One report from Douglas's first days along the Columbia observed that 'Nothing gave me, I think, greater pleasure, than to find *Hookeria lucens* in abundance in the damp, shady forests, growing with a plant whose name also reminded me of another valued friend, the *Menziesia ferrugite*'.[286] In his correspondence to others, Douglas signalled the respect he had for his mentor occasionally referring to him as 'Sir Archibald Menzies' even though he knew no formal knighthood had been bestowed.

News of David Douglas's untimely death in 1834 would have therefore been a horrible shock. Douglas had walked boldly and bravely in Archibald's footsteps, and it was a bitter blow to learn that he had died ascending the peak of Mauna Kea in Hawaii, just thirty-five years old. Meredith Gairdner, the young Hudson Bay surgeon who had travelled with Douglas, and whom Archibald had also entertained at his home prior to his departure, also died in Hawaii, failing to recover from tuberculosis contracted on the Columbia River.

The ultimate devastating blow, two years later, was the loss of Janet. They had met and married late in his life, but it was a true love match. She had made his transition from sailor to civilian possible, provided guidance and direction to his domestic life, given a home to a wandering heart, and had ultimately made him complete. He buried her at nearby Kensal Green Cemetery, the first commercial cemetery in London, which had been developed only a few years earlier. The 22 ha pastoral resting place had plenty of space, but Archibald chose a single plot so that he could lie with her forever, when his final day arrived.[287]

Dawn of Victoria

In addition to outliving so many friends and loved ones, Archibald also lived long enough to see the dusk of the Georgian world that had dominated his life and the dawn of the Victorian era that would equally impress and bewilder him.

The eighteen-year-old queen was crowned in 1837 and soon presided over a host of scientific, technical, economic, and social changes, many of which unfolded before Archibald's eyes. His Edinburgh chemistry professor, Joseph Black, had presented many lectures on latent heat and had been an enthusiastic supporter of James Watt years ago; it was amazing now to see Watt's steam engine technology applied not just to industry but to transportation. Steam-powered ferries began to appear on the Thames, and the London & Greenwich Railway built a short track to transport people from the brand-new London Bridge to the docks at Deptford and Greenwich, covering the distance at an astonishing speed of nearly 32 km/h.

A few kilometres down river from Deptford, where Archibald and Johnstone had once bedazzled Tooweroreo with a battleship tour, a new generation was being amazed by the work being performed at the Blackwall shipyard: they had recently built a steamship for the Hudson's Bay Company – the SS *Beaver* – specifically designed to patrol the Columbia River.[288] Around this time Archibald also heard from his Admiralty contacts that the first screw-propeller ship, the SS *Archimedes*, was being built at the Ratcliffe Cross dockyards, not far from the Blackwall yard. The technology that had increased industrial production was now affecting the navy. Soon, he imagined, ships would sail the world by steam alone.

Archibald was by nature optimistic, and he was perpetually fascinated by discovery and innovation of any kind. He recalled his own youth and the powerful effect a simple bridge and gravel road had made on his community. Nevertheless, he was not unsympathetic to those expressing some concern about the pace and type of advances now taking place and the effect these changes were having on people and communities. The words of Scottish mathematician-philosopher Thomas Carlyle may have struck a chord:

> Our old modes of exertion are all discredited, and thrown aside. On every hand, the living artisan is driven from his workshop, to make room for a speedier, inanimate one. The shuttle drops from the fingers of the weaver, and falls into iron fingers that ply it faster. The sailor furls his sail, and lays down his oar; and bids a strong, unwearied servant, on vaporous wings, bear him through the waters.[289]

The mixed sentiments brought on by such changes may have been best summarised in an 1838 painting by William Turner called *The Fighting Temeraire, tugged to her last berth to be broken up*. It showed the 98-gun HMS *Temeraire*, one of the last battleships active in the Battle of Trafalgar, being towed up the Thames by a paddle-wheel steam tug, on its way to being broken up for scrap. The sun was setting behind the noble, gallant, retired warrior. To some the painting represented a heroic farewell to a valiant, courageous, warrior. But to Archibald it may have suggested to him that the curtain was dropping on the world as he knew it.

Now in his eighties, Archibald was the *Temeraire* of his kind. He was the last of the first to explore the uncharted West Coast of North America, and the last of the first to document to science so many of the plants that feature in parks and gardens around the world today.

Three years after Turner's work was completed, he passed away at home and joined Janet forever at Kensal Green.

Epilogue

The technical, political, commercial, and cultural forces which Archibald observed in the last 40 years of his life gathered momentum in the decades after his death.

The population of London escalated significantly. Fifty years after his death, over six million people were living in the city – five million more than when he began his medical practice – stressing urban life to its limits. In the decade immediately after his death, a series of cholera outbreaks occurred, one traced to a water pump on Broad Street, just half a kilometre away from his old home at Cavendish Square. Five hundred people in the area died in just ten days. Vaccines didn't solve the problem but, thanks largely to the work of Dr John Snow, an effort was made to improve the city's sewer system. Just as Archibald had ordered 'the necessaries' to be removed to the exterior of the *Union*, London developed a sewer system that distanced people and waste from the waterborne cholera and typhoid fever. London's open rivers were covered and put underground.

The face of London above ground also changed significantly in the years immediately following Archibald's death: Nelson's Column and Big Ben – totem poles of a sort – were both completed the year after Archibald died. Tower Bridge would follow 40 years later. The east end would soon turn black in coal dust, and the dockyards which one assembled mighty sail boats began to churn out steamships.

The advances in technology which improved agricultural productivity also turned farmers into factory workers and further accelerated the pace of change. In Scotland these changes also converted clan chiefs from patriarchs to commercial landlords and, although it went against *dùthchas* – the principle that clan members had an inalienable right to rent land in the clan territory – many paid their tenants to emigrate elsewhere like Canada or Australia. The Highland Potato Famine, which particularly affected the Hebrides and the western Highlands in the decade after Archibald's death, caused nearly a third of the population to join them. As author Colin Calloway has observed in his book *White People, Indians, and Highlanders*, 'Highlanders and Indians organized their societies around clan and kinship, occupied land communally as tribal homelands rather than as real estate, and found themselves in the way of an expanding capitalist world that stressed individual ambition, private ownership, and aggressive exploitation of resources for profit'.

On a brighter note, and one that Archibald would have appreciated, Queen Victoria kicked off a resurgence in Highland culture during her reign. Just seven months after Archibald died, the twenty-three-year-old monarch visited his home town of Aberfeldy and was enthusiastically welcomed by clan chief, Sir Neil Menzies, and 50 tenants. They turned out in full Highland dress – which by then was no longer banned.[290] Victoria was so well pleased with the reception, and the area, that in 1855 she requested temporary use of Castle Menzies for Duleep Singh, the last Maharaja of the Sikh Empire and one-time owner of the KohiNoor diamond.

If Archibald was awed by the advances in steam power in his day, he would have been alarmed by its later use and effects in North America. By 1869 the United States had completed a transcontinental railroad, and people on the east coast of the continent no longer had to travel around Cape Horn by ship to get to the West Coast. By that time, people were travelling across land which the United States claimed entirely as its own. George III's 1762 proclamation recognising Indigenous land title west of the original 13 colonies was ignored by the expanding new republic. Much of the new land had been acquired from France by President Jefferson in Archibald's time, but a great deal more was acquired from Spain by President Adams, including all of present-day California. The gold rush there in 1849 provided additional incentive for get-rich-quick travellers. The area of the continent between the Columbia River and Alaska remained in British control, though the Americans claimed that too, based on Captain Gray's commercial activities there during the summer of 1792 when he met with Archibald and Puget near the Strait of Juan de Fuca.

Four years after Archibald's death Great Britain and the United States signed the Treaty of Washington, ceding British claims to the territory between the Columbia River and the 49th parallel. The Hudson's Bay Company, including the hard-working SS *Beaver*, retreated north and moved their headquarters from Fort Vancouver on the Columbia River to Fort Victoria on the southern tip of Vancouver Island. The Colony of British Columbia was established in 1858, partly to halt further American expansion. US President Johnson purchased Alaska from Russia in 1867.

Though the Nootka crisis had spurious origins and an ambiguous ending, there can be little doubt that had Vancouver not come to agreeable terms with Quadra – which is to say, had Quadra refused to leave – the Spanish claims to the area would have remained in place until they were ceded to the United States along with California.

The first governor of the Colony of British Columbia, James Douglas, was well-suited to managing the increasingly multi-ethnic region, and Archibald would probably have got along well with him. He was British but not born in the United Kingdom. He was born, in 1803, in Guyana, to a Creole mother and Scottish father. His wife, Amelia, was the daughter of a French-Irish father and a Cree mother. Douglas was an experienced Hudson's Bay Company man and understood that the company's economic success depended on positive social and business relationships with the Indigenous communities. Like Archibald, Douglas was not the least bit imperial in his approach to languages. Like many who lived and traded in the West Coast prior to the railway and the influx of 20th-century settlers, he was fluent in the Chinook Wawa dialect, a blend of many languages with contributions from Lower Chinook, Nuu-chah-nulth, French, English, some Salishan, and other Indigenous languages. It is estimated that approximately 100,000 people could speak Chinook Wawa in 1875, and it was used in court testimony, newspaper advertising, and everyday conversation.[291]

Isolated from the Dominion of Canada in the east, and sandwiched between the states of Washington to the south and Alaska to the north, the new Colony of British Columbia joined Canada in 1871. A key part of the negotiation was the development of a Canadian transcontinental railway, which connected British Columbia to the eastern provinces in 1885.

Archibald would have been pleased that the area he loved so well had been preserved from American expansionism. But, at the same time, he likely would have been horrified by the speed at which the area was commercialised and developed by people arriving faster than ever by steamship and overland by rail – not as fast as in the United States, but too fast all the same. From 1851 to 1881 the population of British Columbia held steady at about 50,000 people but six short years after the railway was completed it nearly doubled to 98,000. It would double again in the following decade. The volume and speed of emigration over-stressed and drastically changed the delicate balance that had existed between European and Indigenous peoples for the previous 100 years.

Initially, Indigenous peoples had prospered from trade – exporting furs, developing fishing and lumber industries, using metal tools to build more and better canoes, longhouses, art and cultural icons. But the rapid increase in population brought both disease and hubris. The Haida population, which had reached 10,000 by 1850 was reduced by smallpox to just 588 in 1915.[292] The traditional potlach ceremony, which Archibald had happily enjoyed as a guest of Chief Maquinna in 1792, was banned in 1885. At one event, in 1921, twenty Kwakwaka'wakw chiefs were jailed for refusing to sign a court order agreeing not to potlach, and over 600 ceremonial items – masks, regalia, and family heirlooms – were confiscated, and put on display (for a fee) to the general population and many pieces sold to collectors in New York and elsewhere.[293]

As an advocate of George III's 1763 Proclamation, Archibald would have been affronted by the post-Confederation Indian Act, passed by the new Parliament in Ottawa in 1876, which essentially made Indigenous peoples wards of the state and alienated them from direct control over their land, rights, and culture. On the other hand, he would likely have been pleased to learn that early in the 21st century, a female descendant of the Musgamagw Tsawateineuk and Laich-Kwil-Tach people that he met over 200 years earlier at Cape Mudge would be appointed Canada's first Indigenous Attorney General of Canada.[294]

Archibald would likely have been very disappointed by the turn of events in Hawaii. In his lifetime he had every reason to expect Hawaii would continue to have a stronger relationship with Great Britain than with the United States. Even after Kamehameha II died of measles in London, the British influence was strong. The last Queen of Hawaii, Liliuokalani, visited Queen Victoria in London in 1887, and Victoria even agreed to be godmother to Liliuokalani's son, Albert. But Archibald under-estimated the interest Americans would take in the islands, as a conquest first for missionaries, then for commerce, and finally as a conquest for the American republic itself.

In the time since Archibald met Episcopalian clergyman John Howell, in Hawaii in 1794, many more clergymen had followed. In time, many of them also became extensive landowners: Reverend Richard Armstrong, arriving from Pennsylvania in 1832, had some 728 ha under his control by 1850. Sixteen other missionaries around that time averaged 200 ha each. As Archibald had foreseen, Hawaii's climate and geology had huge potential for agricultural production, and it did not take long for missionaries to turn their attention towards more commercial enterprise. Reverend Elias Bond, born in Maine, created the Kohala Sugar Company in 1862. William Shipman, son of a missionary, had acquired 28,000 ha by 1881 and was well on his way to material wealth. James Dole, son of a Unitarian minister, moved to Hawaii in 1899 to start what later became the Dole Food Company. Dole eventually bought the entire island of Lanai, converting it to pineapple production. By 1930, 90% of the world's canned pineapples were produced in Hawaii. Archibald would have been astounded by the levels of agricultural productivity but dismayed and disappointed to see how little this would benefit Indigenous Hawaiians. The effects of disease on the native population – particularly smallpox – would have upset him even more. When Archibald visited the islands, the local population numbered about 300,000 persons but in the decade after his death it was already a fraction of that: about 60,000. By 1920 it was down to around 20,000.[295]

Hawaiian cultural traditions and political systems also became highly stressed in the years following Archibald's death. By 1875 a reciprocity treaty between Hawaii and the United States allowed sugar exports to enter the USA without tariffs and, in exchange, also allowed the US Navy to use Pearl Harbor as a coal station. The increased economic power of the plantation owners decreased the power of the monarchy, and a republic was declared in 1894. Its first President, Sanford B. Dole, the son of missionaries from Maine, and a cousin to James, petitioned the USA to annex the islands, which it did in 1898. Dole's other cousin, Edmund, was appointed attorney general two years later. The reign of Hawaiian monarchs was over, that of the American sugar barons had arrived.

As for his own legacy, there can be no doubt that while Archibald lived a full and accomplished life – indeed, he lived twice as long as average for his era – he was not very good at telling his own story. He saw and experienced the world in a way that few of his own time ever could, progressed from rural Scottish gardener to urban British doctor, and along the way earned many possible epitaphs: navy officer, gentleman, scholar, philosopher, surgeon, medic, physician, herbalist, botanist, ethnologist, artist, cartographer, gardener, adventurer, mountaineer, hunter, marksman, polyglot, and raconteur. But in no way was he ever considered a good publicist.

If Archibald kept a journal during his 1786–88 voyage with Colnett – and it beggars belief that he didn't – no one knows what became of it. Perhaps he destroyed it for reasons we can only guess at. Perhaps he lost it. Perhaps it is hidden away in a distant relative's attic or buried in the archives of a museum. A man with more literary ambition, or perhaps a bigger ego, or both, might have weaved those records into a tale of *Treasure Island* proportions.

The journal Archibald wrote while travelling with Vancouver is extensive and, as has hopefully been revealed in this book, not without literary merit. But Archibald seems to have been more interested in the discoveries and events of today and tomorrow than those of yesterday. His end-of-voyage dispute with Vancouver likely cast a chill over the whole idea of publishing a personal account of the Voyage of Discovery, despite Banks's encouraging him to do so. By the time Vancouver's account was published, completed by his brother Charles, George Vancouver had been in his grave for three years; and Archibald had already turned the page on that chapter of his life and was planning his transition to civilian doctor. Whether he was too modest to tout his own record or too easily distracted by new opportunities and interests is hard to say. Perhaps he was simply more focussed on living a good life than making a name for himself.

One consequence of Archibald's failure to publish his own achievements is that it left the door open for others to publish their accounts first. Despite having collected over 300 plants in the Pacific Northwest, American explorers Lewis & Clark were credited with documenting many plants that Archibald had documented decades earlier. When in 1806 they reached Menzies Island in the Columbia River, so named by Broughton in 1792, they renamed it Canoe Island.[296] It was important to the expanding republic that any British presence in the West Coast be downplayed, or erased if convenient, for the reasons identified earlier in this chapter. Archibald was, in some respect, 'cancelled' 200 years before the practice became popularised in the 2020s. The Pacific dogwood tree which Archibald collected, but failed to name, in 1792 was illustrated and named 40 years later by American ornithologist John James Audubon. In the 1830s Hawaiian missionary Joseph Goodrich claimed no one had climbed Mauna Loa – despite the fact David Douglas stayed with him for three months, and despite the Indigenous people's clear memories of Archibald – and was seemingly happy to be credited for being the first to ascend its summit, although in fact he never did.[297] Others would have continued to trample over his legacy had not the true extent of his botanical achievements been properly acknowledged by his friend William J. Hooker, when he published his work *Flora Boreali-Americana* in 1840. Hooker's book cited nearly 200 plant species attributed to Archibald.[298]

Another issue which limited the renown of Archibald's contributions to natural science, in addition to his failure to quickly publish his journals and name all his plants, was the way his collections were shared. Some seeds and plants went to Kew, some to Banks, some to the Royal Botanic Garden in Edinburgh. Undoubtedly other samples went elsewhere. Likewise, the ethnographic materials he gathered through trade with the Indigenous people he met were provided to different people. Much was provided to Banks, but the list of Archibald's contributions was eventually lost among his papers. Additional artefacts – weapons, baskets, clothing, tools, bowls, blankets – were provided to George III and then passed on to the British Museum without attribution to Archibald. Archibald likely helped his assistant George Hewitt, surgeon's mate on *Discovery*, to collect the nearly 500 objects – arrows, bows, bowls, clubs, fish hooks – from the Pacific Northwest and Hawaii which are currently in the British Museum collections, although Archibald's name is not formally associated with them. It is as though his life's work had been thrown to the four winds.

After his death, Archibald's personal effects were sent to many different places. His personal herbarium was donated to the Royal Botanic Garden, Edinburgh. Some of his correspondence is retained by the Linnaean Society archives, some at the Kew Garden archives, some is elsewhere. The miniature portraits of Archibald and Janet, and his pocket watch, are somewhat surprisingly held by the Royal BC Museum and Archives in Canada.

Despite the failure to publish and promote, and the failure to gather and focus his collections in any clear and attributable way, Archibald was nevertheless well-remembered by those whose lives and interests he touched most. Writing from Hawaii 40 years after Archibald was there, Douglas reported 'I made a journey to the summit of Mowna Roa [Mauna Loa] or the Big or Log mountain which afforded me inexpressible delight … I am not sure whether the venerable Menzies ascended or not … the natives say he did'.[299] In 1892, fifty years after his death, one of the trees Archibald had sent to Kew Garden years before, died. His friend Hooker had it made into a display table containing Archibald's watch, commission on the Vancouver expedition, miniature paintings of Archibald and Janet, and some other papers.[300] In British Columbia, in 1987 – two hundred years after Archibald first visited the area, noted Canadian sculptor Jack Harman was commissioned to create bronze bust of Archibald, along with one of his mentor Carl Linnaeus and his protégé David Douglas. It remains at the VanDusen Botanical Garden, in the city of Vancouver, and currently looks out towards a rose garden and many of the plants and trees of the Pacific Northwest that he so loved. In London, in 2019, one hundred and seventy-seven years after his death, Archibald's Kensal Green gravesite was restored following a fund-raising campaign by local historian Fran Gillespie.

In Scotland, in 2021, Castle Menzies received a copy of the Archibald Menzies bust, created by the original sculptor's son Stephen Harman.[301] It is located, along with some other artefacts, in the castle's 'Archibald Menzies Room'. The bust is placed between two windows: one overlooking the castle garden and one looking toward the ancient standing stones of Croft Moraig.

Sources

Abdulrahman, G. O. Jr., 'John Hunter's (1728–1793) account of venereal diseases', *J Med Biogr.* 24(1) (Feb 2016), 42–4. doi.org/10.1177/0967772013480701. Epub 2014 Jan 30. PMID: 24585621.

Anderson, Bern, 'The Vancouver Expedition: Peter Puget's Journal of the Exploration of Puget Sound May 7–June 11, 1792'. *The Pacific Northwest Quarterly* 30, no. 2 (1939), 177–217, accessed 17 July 2021. jstor.org/stable/41441109.

Andress, David, *1789: The Threshold of the Modern Age* (New York: Farrar, Straus and Giroux, 2008).

Archer, Seth, *Sharks upon the Land: Colonialism, Indigenous Health, and Culture in Hawai'i, 1778–1855* (Cambridge: Cambridge University Press, 2018). doi.org/10.1017/9781316795934.

Atiyah, Michael, 'Benjamin Franklin and the Edinburgh Enlightenment', *Proceedings of the American Philosophical Society* 150, no. 4 (2006), 591–606. jstor.org/stable/4599027.

Bagot, Charles, 'Bygone Days'. *Blackwood's Edinburgh Magazine* 165 (1001), (1899), 461–478. https://www.proquest.com/historical-periodicals/bygone-days/docview/6578891/se-2?accountid=14656.

Bartroli, Tomas, 'Richard Cadman Etches to Sir Joseph Banks: A Plea that Failed'. *British Columbia Historical News*. 8, 3 (1975), 9–18.

Bannerman, John, 'Literacy in the Highlands', in Ian Borthwick Cowan and D. Shaw, eds, *The Renaissance and Reformation in Scotland: Essays in Honour of Gordon Donaldson* (Edinburgh, 1983).

Barnard, Walther M., 'Earliest Ascents of Maua Loa Volcano, Hawai'i'. *Hawaiian Journal of History*, vol., 25 (1991). Available online from: https://evols.library.manoa.hawaii.edu/bitstream/10524/599/2/JL25059.pdf.

Baxby, Derrick, 'Edward Jenner's Unpublished Cowpox Inquiry and the Royal Society: Everard Home's Report to Sir Joseph Banks'. *Medical History* 43, no. 1 (1999), 108–10. https://doi.org/10.1017/S0025727300064747.

Beaglehole, J. C., *The Journals of Captain James Cook* (London: The Hakluyt Society, 1955).

Beresford, William, *A Voyage Round the World; but More Particularly to the North-West Coast of America: Performed in 1785, 1786, 1787, and 1788, in the King George and Queen Charlotte, Captains Portlock and Dixon*, (London: Geo. Goulding, 1789). https://doi.org/10.14288/1.0222775.

Blane, Gilbert, Account of the Battle between the British and French Fleets in the West Indies on the Twelfth of April, 1782, In a Letter to Lord Dalrymple. (UBC Library)

Blane, Gilbert, *A Short Account of the Most Effectual Means of Preserving the Health of Seamen* (1781).

Blane, Gilbert, and Johnston, James, *Letters Etc on the Subject of Quarantine* (London: Philanthropic Reform, 1799).

Blizard, William, *A lecture, On the Situation of the large Blood-Vessels of the Extremities; and the Methods of making effectual Pressure on the Arteries, in Cases of dangerous Effusions of Blood from Wounds: delivered to the scholars of the late maritime school at Chelsea; And first printed for their Use. Third edition. To which is now added, A brief Explanation of the Nature of Wounds, More particularly those received from Fire-Arms* (London: C. Dilly, 1798).

Bodega Y Quadra, *Voyage to the Northwest Coast of America, 1792* (Norman: Arthur H. Clark, 2012).

Bown, Stephen R., *Madness, Betrayal and the Lash: The Epic Voyage of Captain George Vancouver* (Vancouver: Douglas and McIntyre, 2008).

Brockliss, Laurence, Cardwell, John, and Moss, Michael, *Nelson's Surgeon: William Beatty, Naval Medicine, and the Battle of Trafalgar* (Oxford: OUP, 2005), ProQuest Ebook Central, accessed 27 September2021.

Calloway, Colin G., *White People, Indians, and Highlanders: Tribal People and Colonial Encounters in Scotland and America* (New York: Oxford University Press, 2008).

Campbell, Lyle, (1977) *American Indian Languages: The Historical Linguistics of Native America.* (Oxford: Oxford University Press), 396n.34.

Chan, Chelsea, and Demetriades, Andreas K., 'The Contributions of James Carmichael Smyth, Archibald Menzies and Robert Jackson to the Treatment of Typhus in Royal Naval Vessels in the Late 18th Century', *Journal of Medical Biography* (February 2021). https://doi.org/10.1177/0967772021994560.

Constantine, Mary-Ann, and Leask, Nigel, eds, *Enlightenment Travel and British Identities: Thomas Pennant's Tours of Scotland and Wales* (Anthem Press, 2017).

Coote, Jeremy, 'Joseph Bank's Forty Brass Patus', *Journal of Museum Ethnography*, no. 20, (2008), 49–68. www.jstor.org/stable/40793870, accessed 7 July 2021.

Galois, Robert M., *A Voyage to the Northwest Side of America: The Journals of James Colnett, 1786–89* (UBC Press, 2004).

Dixon, *Voyage Around the World.* https://open.library.ubc.ca/collections/bcbooks/items/1.0222775#p6z-6r0f

Dobson, Jessie, *The Surgeons of the Bounty.* https://europepmc.org/backend/ptpmcrender.fcgi?accid=PMC2413561&blobtype=pdf

Douglas, David, 'Sketch of a Journey to the Northwestern Parts of the Continent of North America during the Years 1824–25–26–27', *Quarterly of the Oregon Historical Society* 5, no. 3 (1904), 230–71. http://www.jstor.org/stable/20609621.

Egan, Pierce, *Pierce Egan's Book of Sports, and Mirror of Life* (1832). https://hdl.handle.net/2027/njp.32101064794439?urlappend=%3Bseq=234%3Bownerid=27021597765781045-244, accessed 26 December 2021.

Eltis, David, 'New Estimates of Exports from Barbados and Jamaica, 1665–1701'. *The William and Mary Quarterly* 52, no. 4 (1995), 631–48. https://doi.org/10.2307/2947041.

Emerson, Roger L, 'The Philosophical Society of Edinburgh 1768–1783'. *British Journal for the History of Science* 18, no. 3 (1985), 255–303. http://www.jstor.org/stable/4026382, accessed 11 September 2021.

Fullar, Kate, *The Savage Visit, New World People and Popular Imperial Culture in Britain, 1710–1795* (Berkeley: University of California Press, 2012).

Galois, Bob, 'The Voyages of James Hanna to the Northwest Coast: Two Documents', *BC*

Studies, no 103 (Fall 1994).

Galois, Robert M., *A Voyage to the Northwest Side of America: The Journals of James Colnett, 1786–89* (Vancouver: UBC Press, 2004).

Gilbert, David R., 'Early Exploration and Mapping of the Columbia River', *World Environmental and Water Resources Congress 2014: Water without Borders* © ASCE 2014.

Harvey, A. G., 'Meredith Gairdner, Doctor of Medicine', *British Columbia Historical Quarterly*, Vol IX, No.2 (April 1945).

Hickman, C. (2019), 'The Want of a Proper Gardiner: Late Georgian Scottish Botanic Gardeners as Intermediaries of Medical and Scientific Knowledge'. *British Journal for the History of Science*, 52(4), 543–567. https://doi.org/10.1017/S0007087419000451.

Hill, Beth, and Converse, Cathay, *The Remarkable World of Frances Barkley 1769-1845* (Touchwood Editions, 2003).

Hellman, Lisa, 'Olof Lindahl and the 1770s and 1780s'. In: *This House Is Not a Home* (Leiden: Brill, 2018). https://doi.org/10.1163/9789004384545_008.

Home, Everard, and Menzies, Archibald, *A Description of the Anatomy of the Sea Otter* (London: Royal Society of London, 1796). Eighteenth Century Collections Online, https://link.gale.com/apps/doc/CW0108305327/ECCO?u=ubcolumbia&sid=bookmark-ECCO&xid=46406128&pg=5, accessed 22November 2021.

'Clarice B. Taylor's Tales about Hawaii, Education of Kualelo in Plymouth England', *Honolulu Star Bulletin* (26 March 1953), 30. https://www.newspapers.com/clip/40645045/honolulu-star-bulletin/, accessed 4 August 2021.

Howard, Richard A., 'The St. Vincent Botanical Garden – The Early Years', *Harvard Papers in Botany* 1, no. 8 (1996), 1–6. http://www.jstor.org/stable/41761509.

Howay, F. W., 'Some Notes on Cook's and Vancouver's Ships, 1776-80, 1791-95'. *Washington Historical Quarterly* 21, no. 4 (1930), 268–70. http://www.jstor.org/stable/40475364, accessed 17 July 2021.

Howe, K. R. 'Review Article: The Making of Cook's Death' *Journal of Pacific History* 31, no. 1 (1996), 108–18. http://www.jstor.org/stable/25169289.

Hubbell, Alvin A., *A cataract knife of excellent shape and proportion devised a century and a half ago, by Dr. Thomas Young, of Edinburgh, and the knives which preceded it* (Buffalo: The Ophthalmic Record, 1900). https://wellcomecollection.org/works/h35decve/items?canvas=7, retrieved September 15, 2021.

Hull, Gillian, 'Archibald Menzies (1754–1842): A Respected Surgeon/Naturalist'. *Journal of Medical Biography* 9, no. 4 (November 2001), 226–30. https://doi.org/10.1177/096777200100900407.

Jewitt, John Rodgers (1896), *The adventures of John Jewitt*, ed. Robert Brown (London: Clement Wilson, 1899). Retrieved from: https://open.library.ubc.ca/collections/chung/chungpub/items/1.0114653#p7z-5r0f.

Judd, Bernice, 'Index of Vessels and Persons'. In: *Voyages to Hawai'i Before 1860: A Record, Based on Historical Narratives in the Libraries of the Hawaiian Mission Children's Society and the Hawaiian Historical Society, Extended to March 1860*, ed. Helen Yonge Lind, 45–96 (University of Hawai'i Press, 1974). https://doi.org/10.2307/j.ctvp2n4t4.7.

Justice, Clive L., *Mr. Menzies' Garden Legacy, An Illustrated History* (Vancouver: Big Leaf Maple Books, 2017).

Keevil, J. J., 'Archibald Menzies, 1754–1842', *Bulletin of the History of Medicine*, vol. 22, no. 6, (1948), 796–811. www.jstor.org/stable/44442236, accessed 13 July 2021.

Lamb, Kaye W., *The Voyage of George Vancouver, 1791–1795*, Volumes I-IV.

Lepore, Jill, 'Goodbye Columbus', *New Yorker* (30 April 2006). https://www.newyorker.com/magazine/2006/05/08/goodbye-columbus, retrieved 26 September 2021.

Mathes, Valerie Sherer, 'Wickaninnish, a Clayoquot Chief, as Recorded by Early Travelers'. *Pacific Northwest Quarterly 70*, no. 3 (1979), 110–20. http://www.jstor.org/stable/40489855.

McCarthy, James, *Monkey Puzzle Man*. (Dunbeath: Whittles Publishing, 2008).

McLaren, Jennifer, 'Celebrating the Battle of the Saintes: Imperial News in England and Ireland, 1782'. *Éire-Ireland*, Vol 51, No 1 & 2, (Spring/Summer 2016), 34–60.

Menzies, Archibald, *Hawaii Nei 128 Years Ago* (Honolulu: 1920). https://ia601603.us.archive.org/20/items/hawaiineiyearsa00wilsgoog/hawaiineiyearsa00wilsgoog.pdf.

Menzies, Archibald, 1790–1794. *Journal of Archibald Menzies, Surgeon and Botanist on Board Discovery* (1794). Available through: Adam Matthew, Marlborough, Empire Online, https://www.empire.amdigital.co.uk/documents/detail/4006406.

Menzies, Archibald, *Menzies' Journal of Vancouver's Voyage April to October 1792*, ed. C. F. Newcomb, MD. (Victoria, BC: Legislative Assembly, 1923).

Menzies, D. P., *The Red and White Book of Menzies: The History of the Clan Menzies and its Chiefs* (Glasgow: Banks and Co., 1894).

Moeller, Beverley B., 'Captain James Colnett and the Tsimshian Indians, 1787', *Pacific Northwest Quarterly* 57, no. 1 (1966), 13–17. http://www.jstor.org/stable/40488086, accessed 16 August 2021.

Moser, Chas, *Reminiscences of the west coast of Vancouver Island* (Victoria: Acme Press, 1926). Retrieved from UBC library: https://open.library.ubc.ca/collections/bcbooks/items/1.0354345#p173z-4r0f:frances.

Musgrave, Toby, *The Multifarious Mr Banks: From Botany Bay to Kew, The Natural Historian Who Shaped the World* (Yale University Press, 2020).

Murray, David F., 'The Alaska Travel Journals of Archibald Menzies, 1793–1794, with an Introduction and Annotation by Wallace M. Olson'. *Arctic* 47 (3) (1994), 207–319. https://doi.org/10.14430/arctic1479.

Naish, John, *The Interwoven Lives of George Vancouver, Archibald Menzies, Joseph Whidbey and Peter Puget: The Vancouver Voyage of 1791–1795*. (Lewiston: Edwin Mellen, 1996).

Naish, J., 'Archibald Menzies: Surgeon botanist' *West Engl Med J* (Dec 1991) 106 (4), 108–9. PMID: 1820080; PMCID: PMC5115067.

Nolte, H. J., *John Hope (1725–1786): Alan G Morton's Memoir of a Scottish Botanist* (Edinburgh: Royal Botanic Garden, 1986).

Pennant, Thomas, *A Tour in Scotland, and Voyage to the Hebrides, 1772* (Cambridge: Cambridge University Press, 2014). https://doi.org/10.1017/CBO9781107589384.

Paine, Thomas, *The Rights of Man*.

Riedel, Stefan, 'Edward Jenner and the history of smallpox and vaccination', *Proceedings* (Baylor

University. Medical Center) vol. 18,1 (2005), 21–5. https://doi.org/10.1080/0899 8280.2005.11928028.

Reid, Joshua, *The Sea Is My Country: The Maritime World of the Makahs, an Indigenous Borderlands People* (Yale University Press, 2015). retrieved at:

Restarick, Rev. Henry B., 'The First Clergyman Resident in Hawaii: Thirty-Second Annual Report of the Hawaiian Historical Society for the Year 1923 with Papers Read at the Annual Meeting, January 24 1924'.

Shulman, Stanford T., Shulman, Deborah L., Sims, Ronald H., 'The Tragic 1824 Journey of the Hawaiian King and Queen to London', *Pediatric Infectious Disease Journal.* Vol 28, No. 8 (August 2009), https://citeseerx.ist.psu.edu/pdf/932eb91 11436c09b227759e067550932acbc446e, retrieved December 28, 2021, retrieved 1 April 2024.

Smith, Adam, *The Wealth of Nations* [1776].

Smyth, James Carmichael, *An account of the experiment made at the desire of the Lords Commissioners of the Admiralty, on board the Union hospital ship, to determine the effect of the nitrous acid in destroying contagion, and the safety with which it may be employed. In a letter addressed to the Right Hon. Earl Spencer, &c. &c. &c. By James Carmichael Smyth, M. D. F.R.S., Fellow of the Royal College of Physicians, and Physician Extraordinary to His Majesty, published with the approbation of the lords commissioners of the Admiralty* (London: J. Johnson, 1796). https://go.gale.com/ps/i.do ?p=ECCO&u=ubcolumbia&id=GALE|CW 0108247382&v=2.1&it=r&sid=bookmark-ECCO&sPage=1&asid=7f372d5f, accessed 22 Nov. 2021.

Sutherland, David, 'Halifax Merchants and the Pursuit of Development, 1783–1850'.

Canadian Historical Review, Volume 59 Issue 1 (March 1978), 1–17.

The history of the feuds and conflicts among the clans in the northern parts of Scotland and in the western isles; from the year M.XXXI. unto M.DC.XIX. Now first published from a manuscript, wrote in the reign of King James VI. Glasgow: printed by Robert and Andrew Foulis, M.DCC.LXIV. [1764]. https:// link.gale.com/apps/doc/CW0102148258/ ECCO?u=ubcolumbia&sid=bookmark-ECCO&xid=b3971e08&pg=1.

Tolstoy, Nikolai, *The Half-Mad Lord: Thomas Pitt, 2nd Baron Camelford (1775–1804)* (Ebenezer Baylis and Son, 1978).

Tulchinsky T. H. 'John Snow, Cholera, the Broad Street Pump: Waterborne Diseases Then and Now'. In: *Case Studies in Public Health* (Academic, 2018), 77–99. https://doi.org/10.1016/ B978–0-12-804571–8.00017–2.

Vancouver, George, *A Voyage of Discovery to the North Pacific Ocean and Round the World.* (London: G. G. and J. Robinson; J. Edwards, 1798).

Wystrach, V.P., 'Anna Blackburne (1726–1793): A Neglected Patroness of Natural History'. *J. Soc. Biblphy Nat Hist* 8 (2) (1977), 148–168.

ENDNOTES

1 Possibly a reference to Christian scripture. Revelations 6:8 of the Bible: 'And I looked, and behold a pale horse: and his name that sat on him was Death, and Hell followed with him. And power was given unto them over the fourth part of the earth, to kill with sword, and with hunger, and with death, and with the beasts of the earth'.

2 The word 'clan' is derived from the Scottish Gaelic word clann meaning 'children, offspring, descendants'.

3 While mostly patriarchal, the clan system could also be matriarchal – it allowed a woman to be chief. Similarly, while the chiefdom was mostly hereditary, it was not necessarily so. The Scottish clan system was an open and democratic form of centralized authority.

4 Maps which show well-defined clan territories with clear boundaries are a Victorian invention; in Archibald's youth the tribal boundaries, like those he would encounter years from now in places far away, were more likely sensed than seen.

5 Sir Robert and the castle gardens: D. P. Menzies, DP, FSA Scot., The Red and White Book of Menzies: The History of the Clan Menzies and its Chiefs (Glasgow: Banks & Co., 1894).

6 Castle Menzies gean (cherry) trees Clan Menzies: Menzies, The Red and White Book of Menzies, 457.

7 The first larch trees in Great Britain: Menzies, The Red and White Book of Menzies, 456.

8 George III reading materials and notes: 'The Georgian Papers', https://georgianpapers. com/2017/01/19/farmer-georges-notes-agriculture/, accessed 20 August 2021 .

9 Two hundred years later, in 1968, a group of musicians led by Scottish flautist frontman Ian Anderson adopted Jethro Tull's name for their band.

10 Merino wool: Royal Collection Trust, https://www. rct.uk/collection/1057006/a-first-account-of-the-spanish-merino-flock-of-his-majesty-george-iii, accessed 20 August 2021.

11 The physic garden is now long gone, but there is a historical marker on platform 11 at Edinburgh Waverley [railway] Station marking where it once was.

12 John Campbell of Glenlyon was the son of Robert Campbell (1632–1696), infamous for his role in the 13 February 1692 Massacre of Glencoe. The anecdote regarding John's taking of Castle Menzies

in 1716 was related to the author in written correspondence with local historian Tommy Pringle, 28 August 2021.

13 This and other excerpts from The Statistical Account of the Parish of Dull, written by Rev. Archibald Menzies, is reproduced in: Menzies, The Red and White Book of Menzies, 412.

14 Voltaire quote and additional context can be found here: https://www.scottishreviewofbooks.org/free-content/voltaire-versus-lord-kames-and-the-need-for-a-soundbite/, retrieved 25 July 2021.

15 H. J. Nolte, John Hope (1725–1786): Alan G Morton's Memoir of a Scottish Botanist (Royal Botanic Garden: Edinburgh, 1986), 52.

16 Nolte, John Hope, 58.

17 Nolte, John Hope, 86.

18 Nolte, John Hope, 38.

19 Email correspondence with Danielle Spittle, CRC (Edinburgh University Library), 16 June 2021. 'Archibald Menzies studied Medicine at the University of Edinburgh for 8 academic years between 1771 and 1780, but didn't graduate. This was quite common: graduating was expensive, and most employers didn't require a degree. Because Archibald didn't graduate, the only student records that we have for him are his signatures in the matriculation (enrolment) album. The matriculation entries tell us the subjects that Archibald studied and the names of his professors. Interestingly, an Archibald Menzies also matriculated in Arts in 1776–77 and 1777–78, studying Maths and Natural Philosophy (i.e. Science). Because we only have signatures with no other identifying information such as addresses, we can't say for certain whether the Archibald Menzies who matriculated in Arts is the same person as the Archibald Menzies who matriculated in Medicine. However, to my eyes the signatures look very similar, so it's my opinion that "your" Archibald also studied Maths and Natural Philosophy alongside his medical studies.'

20 The term Secundus (second) helps distinguish Alexander Monro from his father Alexander Monro Primus (first) who was also a professor of anatomy at the University of Edinburgh. Primus was the son of John Monro, a surgeon who helped establish the

medical school at the University of Edinburgh in 1726 – now the oldest medical school in the UK.

21 From: Robert Hamilton and William Cullen, 'Memorial by Robert Hamilton, Professor of Anatomy and Botany, and Dr William Cullen, Professor of Medicine, to the University concerning the planting of more trees and shrubs in the College garden in place of the decayed fruit trees', Glasgow University Archives, GUA 5412.

22 The bridge was completed in 1772; the first Waverley train station would not be built until 1846.

23 It is not proven that Hope ever invited Archibald to join him at an Aesculapian Club meeting, but the author thinks, given their relationship, it is probable.

24 https://parliamentssquareedinburgh.net/edinburgh-clubs-and-societies/, retrieved 15 September 2021.

25 Emerson, Roger L. 'The Philosophical Society of Edinburgh 1768–1783'. The British Journal for the History of Science 18, no. 3 (1985), 275, http://www.jstor.org/stable/4026382, accessed 11 September 2021.

26 The other founders were Maclaurin, Lord Hope, Andrew Plummer, and Alexander Lind of Gorgie.

27 The letter from Lord Morton can be seen and read at https://nla.gov.au/nla.obj-223065583/view, retrieved 16 September 2021. Unfortunately, while Cook was a master navigator, he was neither a philosopher nor a miracle worker, and unable to fully embrace Lord Morton's hints. In taking possession of the east coast of Australia in the name of the British Crown, Cook contradicted the advice offered by Morton. By doing so, it may be argued he showed his primary duty was to the colonial ambitions of the British government and of the Admiralty.

28 Full text of the Proclamation of 1763: https://avalon.law.yale.edu/18th_century/proc1763.asp, retrieved 19 September 2021.

29 Michael Atiyah, 'Benjamin Franklin and the Edinburgh Enlightenment', Proceedings of the American Philosophical Society 150/4 (2006), 591, http://www.jstor.org/stable/4599027, retrieved 16 September 2021.

30 Jeremy Coote, 'Joseph Banks's Forty Brass Patus', Journal of Museum Ethnography, no. 20 (2008), 49–68. JSTOR, www.jstor.org/stable/40793870, accessed 7 July 2021.

31 When James Cook first encountered the villagers at Yuquot in 1778, they directed him to 'come around' (Nuu-chah-nulth nuutkaa is to 'circle around') with his ship to the harbour. Cook interpreted this as the First Nation's name for the inlet now called Nootka Sound. The term was also applied to the Indigenous inhabitants of the area.

32 Village of Tahsis website, http://villageoftahsis.com/wp-content/uploads/2016/06/NootkaSoundhistory-1.pdf, 8 September 2021.

33 The first documented ascent of Ben Nevis was performed by James Robertson on 19 August 1771. Robertson was a student of John Hope and worked at the Royal Botanic Garden Edinburgh. The exact date of Archibald's ascent remains unknown to the author but, regardless, Archibald would have been among the first trained botanists to make the ascent.

34 William Pitcairn's fifteen-year-old nephew, Robert, was a midshipman on HMS Swallow in 1767 when he spotted a remote Pacific island which was subsequently named after him: Pitcairn Island.

35 Banks's comment on Fothergill's estate, http://www.e7-nowandthen.org/2018/11/john-fothergill-1712-1780-quaker.html, retrieved 21 September 2021.

36 Some literature contends Archibald received his commission later (after being in Wales) but according to this report, which the author thinks more likely accurate, Archibald received his commission on 8 October 1778, having served as surgeon's mate on hospital staff before being commissioned (before he went to Wales): William Johnston (Col). Roll of the commissioned officers in the medical service of the British Army (1917), 51.

37 The historical record of Archibald's time in Wales is mostly anecdotal. Most biographical sketches, if they mention these years at all, say he was 'assistant to a surgeon', though more likely he was 'Assistant Surgeon' – a term and role commonly used in the Royal Navy at this time. But which surgeon he assisted, or where precisely he was an assistant surgeon, remains a mystery.

38 Laurence Brockliss, John Cardwell, and Michael Moss, Nelson's Surgeon: William Beatty, Naval Medicine, and the Battle of Trafalgar (Oxford: Oxford University Press, 2005), 10. ProQuest Ebook Central, accessed 27 September 2021.

39 Meaning each cannon ball weighed 11 kg.

40 This was the Battle of Dogger Bank. Twelve British ships engaged 13 Dutch ships, scoring a slight victory for the British. The battle was part of the Fourth Anglo-Dutch War, but was not unconnected to the Dutch role in shipping French supplies to America for the revolution there.

41 Most biographical sketches maintain that Archibald served as 'Assistant Surgeon' on Nonsuch and as 'Surgeon' on all ships he served on thereafter. However, the author has not been able to find any primary sources to confirm that he was assistant surgeon on Nonsuch. Rather, the author believes it more likely Archibald was engaged as full surgeon on Nonsuch given that he had already been a surgeon's mate before receiving his Board qualification in 1778 and had also served two years as 'Assistant Surgeon' in Wales. The author contends Archibald was well qualified for an appointment as full surgeon by the time he joined Nonsuch.

42   Washington's correspondence and more
      information about the naval aspect of the American
      Revolutionary War: 'The American Revolution
      at Sea', https://www.americanrevolutioninstitute.
      org/exhibition/the-american-revolution-at-sea/,
      retrieved 30 September 2021.

43   'The American Revolution at Sea'.

44   Gilbert Blane, A Short Account of the Most
      Effectual Means of Preserving the Health of Seamen
      [introduction] (1781), https://quod.lib.umich.edu/e/
      ecco/004778092.0001.000/1:5?rgn=div1;view=fullte
      xt, retrieved 30 September 2021.

45   Charles Douglas's 1727 birthplace is unknown but
      was likely not far from Edinburgh, where he died
      in 1789.

46   Gilbert Blane. Account of the Battle between the
      British and French Fleets in the West Indies on
      the Twelfth of April, 1782, In a Letter to Lord
      Dalrymple.

47   Blane, Account of the Battle.

48   Jennifer McLaren, 'Celebrating the Battle of the
      Saintes: Imperial News in England and Ireland,
      1782', Irish–American Cultural Institute, Vol 51, No
      1 & 2 (Spring/Summer 2016), 39.

49   Jennifer McLaren, 'Celebrating the Battle of the
      Saintes'.

50   Blane, Account of the Battle.

51   James Vashon's ships and shipmates: https://
      www.vashonhistory.com/Publications/
      Commentaries/capt_james_vashon_bio.pdf,
      retrieved 2 October 2021.

52   Jennifer McLaren, 'Celebrating the Battle of the
      Saintes', 39–40.

53   The British would not succeed in abolishing the
      slave trade until the Slavery Abolition Act was
      passed in 1833.

54   According to this 1797 publication, Halifax was
      home to 4,000 people and 700 houses in 1793. It
      would be reasonable to assume the population
      and number of houses was much smaller when
      Archibald arrived in 1782–3: Jedidiah D. D. Morse,
      The American Gazetteer (Boston: S. Hall and
      Thomas & Andrews, 1797), 228, https://archive.
      org/stream/americangazettee00mors#page/n227/
      mode/2up.

55   Letter from Archibald Menzies to John Hope, May
      30, 1784. From: James McCarthy, Monkey Puzzle
      Man (Dunbeath: Whittles, 2008), 44–45.

56   David Andress, 1789: The Threshold of the Modern
      Age (New York: Farrar, Straus and Giroux, 2008), 123.

57   In December 1775, after hearing of Lord Dunmore's
      proclamation, George Washington, commanding
      the Continental Army in Cambridge, is reported to
      have said of his slaves 'There is not a man of them
      but would leave us if they believed they could make
      their escape'.

58   Letter from Archibald Menzies to John Hope, July 4,
      1785. From: James McCarthy, Monkey Puzzle Man, 47.

59   Letter from Archibald Menzies to John Hope, 21
      August 1786. From: McCarthy, Monkey Puzzle
      Man, 54.

60   Present-day currency estimates have been obtained
      by using the following calculator and converting
      the GBP value to US dollar value at time of writing:
      Currency converter used to find present day values:
      https://www.nationalarchives.gov.uk/currency-
      converter/#currency-result.
         Another report said that sailors on Cook's voyage
      had bought skins at Nootka 'which did not cost the
      purchaser six-pence sterling [and] sold in China for
      100 dollars.' Beverly B. Moeller, 'Captain Colnett and
      the Tsimshian Indians, 1787', The Pacific Northwest
      Quarterly, Vol 57, No. 1 (January 1966), 13.

61   The licence was granted in August of 1785, while
      Archibald was still sailing between the Caribbean
      and Halifax.

62   J. J. Keevil, Archibald Menzies, 1754–1842,
      Bulletin of the History of Medicine, vol. 22, No. 6
      (November–December, 1948), 798.

63   Richard Etches's letter to Banks after meeting
      Archibald: Archibald's instructions/meeting with
      Etches: Bartroli, Tomas. 'Richard Cadman Etches
      to Sir Joseph Banks: A Plea that Failed'. British
      Columbia Historical News 8, 3 (1975), 9–18.

64   An account of Archibald's visit to Banks's herbarium
      at this time can be found in Keevil, Archibald
      Menzies, 798. Banks's home collection preceded
      London's National History Museum by a century.

65   Daniel Solander died at Banks's home on 13 May 1782.

66   Instructions to Colnett and his mission as a
      commercial voyage: Robert M. Galois, A Voyage
      to the Northwest Side of America: The Journals of
      James Colnett, 1786–89 (UBC Press, 2004), 11.

67   Andrew Taylor's journal comments regarding
      ducking: Galois, A Voyage to the Northwest Side of
      America, 81.

68   Joseph Banks's journal entry of 25 October 1768
      recounts the ducking ceremony: Joseph Banks, The
      Endeavour Journal of Sir Joseph Banks 1768–1771
      (Sydney: University of Sydney Library, 1997), 48.
      http://setis.library.usyd.edu.au/ozlit/
         © University of Sydney Library, State Library of
      NSW. The texts and Images are not to be used for
      commercial purposes without permission.

69   Probably at a location now known as Colnett Bay.
      On the south side of the Island is another smaller
      island called Menzies Island, undoubtedly named
      after Archibald.

70   No one would know until much later that the
      scheme wouldn't last more than a year or two in this
      harsh, rocky, and remote location.

71   Colnett's journal remarks regarding his plan to

refresh supplies at the Gallegos Islands (instead of Hawaii): Galois *A Voyage to the Northwest Side of America*, 94.

72 Colnett journal remarks regarding the Spanish map for Gallegos: Galois *A Voyage to the Northwest Side of America*, 94–95.

73 Blane's pamphlet was published in 1780. Later, as Commissioner of the Sick and Wounded Board, Blane would persuade the Admiralty to go against the theories of the medical establishment and introduce lemon juice as daily addition to the naval diet in 1795. Vitamin C itself would not be discovered until 1928.

74 Colnett's journal description from McCarthy, *Monkey Puzzle Man*, 58.

75 Tonnage is a measure of the cargo-carrying capacity of a ship. The term derives from the taxation paid on tuns or casks of wine. In modern maritime usage, 'tonnage' specifically refers to a calculation of the volume or cargo volume of a ship. A ton is about 98% of a metric 'tonne'.

76 Beth Hill and Cathay Converse, *The Remarkable World of Frances Barkley 1769–1845* (Touchwood Editions, 2003), 50.

77 Strange, as his full name hints, was the godchild of Bonnie Prince Charlie. His father, Robert, originally from Orkney Island in Scotland, was a veteran of the Jacobite Rebellion of 1745.

78 Menzies' journal, September 5, 1792. From McCarthy, *Monkey Puzzle Man*, 140.

79 Wynee was probably an attempted spelling of 'wahine', the Hawaiian word for 'woman'. She sailed with Barkley to Nootka, then to Macao where she transferred to John Meares's ship en route to Hawaii. Sadly, she died of fever before reaching home and was buried at sea. She was the first Hawaiian to visit British Columbia.

80 Frances Barkley was the first European woman to set foot in British Columbia.

81 After his time in Nootka, Barkley sailed south and met Maquinna's rival, Chief Wickaninnish of the Tla-o-qui-aht people. When the Barkleys left Nootka in mid-July, Frances wrote in her diary, 'A day or two after sailing from King George's Sound we visited a large sound in latitude 49.20 North, which Captain Barkley named Wickanninish's Sound [Clayoquot Sound] the name given being that of a chief who seemed to be quite as powerful a potentate as Maquilla at King George's Sound. Wickaninnish has great authority and this part of the coast proved a rich harvest of furs for us'. A little further south, just past present-day town of Ucluelet, Barkley noted another large sound which he simply named Barkley Sound.

82 Colnett's journal entry from Galois, *A Voyage to the Northwest Side of America*, 108.

83 Archibald's letter to Banks, dated 4 April 1790, responded to an enquiry from Banks asking his advice on what sort of trade items to include on the Discovery mission. The short letter is reproduced in: Galois, *A Voyage to the Northwest Side of America*, 60.

84 Samples of abalone shells from this era can be viewed at the University of British Columbia's Museum of Anthropology in Vancouver, BC.

85 Letter from Archibald to Banks regarding the Pata-patoo, 14 July 1789. From McCarthy, *Monkey Puzzle Man*, 66.

86 Oughomeize was a Muchalaht chief on the eastern shores of Nootka Sound, possibly based at the villages of Cheesish or Mooya.

87 Galois, *A Voyage to the Northwest Side of America*, 104.

88 Mr Elliot's letter can be read at Bob Galois, 'The Voyages of James Hanna to the Northwest Coast, Two Documents'. *BC Studies*, no 103 (Fall 1994), 85–86.

89 This account is from July 1787, at about the time Archibald, Colnett, and Duncan were arriving on the scene. The original text uses 'f's where we nowadays use 's's. These have been changed for ease of reading. From William Beresford, *A voyage round the world; but more particularly to the north-west coast of America: performed 1785, 1786, 1787, and 1788, in the King George and Queen Charlotte, Captains Portlock and Dixon*, 208. UBC Library https://open.library.ubc.ca/collections/bcbooks/items/1.0222775#p266z-4r0f, retrieved July 2021.

90 William Beresford, *A voyage round the world*, 267–268. https://open.library.ubc.ca/collections/bcbooks/items/1.0222775#p266z-4r0f, retrieved July 2021.

91 This Indigenous perspective of first contact was recorded by William Benyon on 9 February 1916 during field work at Gitxaala, where he interviewed sixty-year-old Ganhada Chief George McCauley. This account, and other versions telling the Indigenous perspective of first contact, can be found in Galois, *A Voyage to the Northwest Side of America*, 268–272.

92 Joshua Reid, *The Sea Is My Country: The Maritime World of the Makahs, an Indigenous Borderlands People* (Yale University Press, 2015), 64.

93 Elliot wrote of encountering a war canoe 'adorned round the gunwale with three rows of human teeth': Galois, *The Voyages of James Hanna to the Northwest Coast*, 87.

94 Archibald may have gathered the arbutus seed for Sir Robert on his later voyage with Vancouver. But there really is an arbutus tree near Castle Menzies that dates to the late 1700s.

95 Known as the Sandwich Islands at the time but referred to here as Hawaii for ease of understanding.

96 Galois, *The Voyages of James Hanna to the Northwest Coast*, 88.

97   Galois, A Voyage to the Northwest Side of America, 61.

98   One of de Bougainville's crew was Jean Baré, the first woman to circumnavigate the globe.

99   It is commonly reported that Cook's crew introduced venereal disease to the Hawaiian Islands. This article offers a compelling argument that the disease was likely already present when Cook arrived· G. O. Abdulrahman Jr., 'John Hunter's (1728–1793) account of venereal diseases', J Med Biogr. 24(1) (Feb 2016), 42–4, https://doi.org/10.1177/0967772013480701. Epub 2014 Jan 30. PMID: 24585621, page 2.

100   Galois, A Voyage to the Northwest Side of America, 195.

101   Colnett's journal on 23 February 1788 notes that Archibald 'had taken a good deal of pains to make himself acquainted with the language'. McCarthy, Monkey Puzzle Man, 62.

102   Journal entry, 14 March 1793. Archibald Menzies, Hawaii Nei 128 Years Ago (Honolulu, T.H., 1920), 109, https://ia902605.us.archive.org/10/items/hawaiineiyearsa00wilsgoog/hawaiineiyearsa00wilsgoog.pdf.

103   Journal entry, 14 March 1793. Menzies, Hawaii Nei 128 Years Ago, 110.

104   Journal entry, 14 March 1793. Menzies, Hawaii Nei 128 Years Ago, 109.

105   Galois, A Voyage to the Northwest Side of America, 188.

106   Galois, A Voyage to the Northwest Side of America, 366n57.

107   Galois, A Voyage to the Northwest Side of America, 18.

108   Present-day currency estimates have been obtained by using the following calculator and converting the GBP value to US dollars at time of writing: Currency converter used to find present day values: https://www.nationalarchives.gov.uk/currency-converter/#currency-result.

109   These were not factories in the modern sense of the word but more like a warehouse or trade office.

110   W. Kaye Lamb and Tomas Bartroli, 'James Hanna and John Henry Cox: The First Maritime Fur Trader and His Sponsor'. BC Studies, no 84 (Winter 1989–90), 31.

111   Lamb and Bartroli, 'James Hanna and John Henry Cox', 32.

112   Galois, A Voyage to the Northwest Side of America, 308n154.

113   Anderson died on 3 August 1778.

114   Lamb and Bartroli, 'James Hanna and John Henry Cox', 28.
      Whether the Otter was lost in a storm or if the crew were the victim of attack remains a mystery. This pattern of danger would continue in the area, long after Archibald left for the last time, some even suggesting the area held a curse for surgeons: the next ship's doctor to visit the Pacific Northwest after Archibald, a Doctor White, came to Astoria (Fort George, in what is now the State of Oregon) some 30 years later, in 1811, and 'became suddenly deranged, jumped overboard, and was drowned'. Shortly afterwards, a Doctor Crowly, who came from Edinburgh to Astoria as medical officer to the North West Company, shot a man, was charged with murder and sent home to stand trial. Mr. Downie, surgeon to the Company's ship Colonel Allan, of London, committed suicide by shooting himself in his cabin while off Fort George in 1816. All this to underscore the point that the role of surgeon/adventurer was not for the weak of body or mind. Archibald's voyage would have been a triumph had he merely survived it.

115   'French Monarchy Overthrown: King and Family Imprisoned – archive, 1792'. The Guardian (10 August 2022), https://www.theguardian.com/world/2022/aug/10/french-monarchy-overthrown-king-louis-imprisoned-1792?CMP=share_btn_tw, retrieved 12 August 2022.

116   Letter from Archibald to Banks, 14 July 1789. From: McCarthy, Monkey Puzzle Man, 66.

117   Letters to and from Smith, in English, French, Latin, etc are preserved in the archives of the Linnean Society and can be viewed online at http://linnean-online.org/view/correspondence_by_date/smith=5Fcorrespondence/.

118   Tooworero is sometimes also known in other works as 'Kualelo' because modern day writers think that may be a more appropriate modern-day spelling of his spoken name. But the name used by Archibald in his journals is Tooworero. He was also the first Hawaiian to visit London.

119   Menzies, Hawaii Nei 128 Years Ago, 18–19.

120   Rutherford would be celebrated later in life for being the discoverer of nitrogen, and for being the uncle of Sir Walter Scott.

121   Some have said that Archibald's older brother William worked at the RBGE, but the author was unable to confirm it. He was, however, able to confirm that Archibald's brother Robert worked there. Direct email correspondence between the author and RBGE Archives, 27 July 2021.

122   Pitcairn's garden: https://history.rcplondon.ac.uk/inspiring-physicians/william-pitcairn, retrieved 6 August 2021.

123   Letter from Archibald Menzies to Banks, 8 October 1789. From: McCarthy Monkey Puzzle Man, 78.

124   Tomas Bartroli, 'Richard Cadman Etches to Sir Joseph Banks: A Plea that Failed'. British Columbia Historical News. 8, 3 (1975), 9–18.

125 M. Tovell Freeman, Robin Iglis, and Iris H.W. Engstrand, Voyage to the Northwest Coast of America: 1792 Juan Francisco de la Bodega y Quadra and The Nootka Sound Controversy, 29.

126 This and additional colour commentary by Bligh on Huggins's alcoholism: Jessie Dobson, 'The Surgeons of The Bounty', 1. https://europepmc.org/backend/ptpmcrender.fcgi?accid=PMC2413561&blobtype=pdf, retrieved 16 August 2021.

127 Bligh, determined that Nelson's loyalty not be forgotten, named Mount Nelson, in Tasmania, after him.

128 Remarkably, the Discovery's Arnold Chronometer still survives and can be viewed at the Vancouver Maritime Museum in Vancouver, BC. Lt. Baker's telescope – perhaps the one he used to spy Mount Baker – is also held at this museum, as is his sextant.

129 'Draught of Instructions for Mr. Menzies (Banks papers)'. Victoria University of Wellington Library. Historical Record of New Zealand, http://nzetc. victoria.ac.nz/tm/scholarly/tei-McN01Hist-t1-b4-d5.html, retrieved 20 October 2021.

130 J. J. Keevil, 'Archibald Menzies, 1754–1842.' Bulletin of the History of Medicine, vol. 22, no. 6, (1948), 796–811. JSTOR, www.jstor.org/stable/44442236, 799–800.

131 F. W. Howay, 'Some Notes on Cook's and Vancouver's Ships, 1776–80, 1791–95'. The Washington Historical Quarterly 21, no. 4 (1930), 268–70, accessed 17 July 2021.

132 Archibald Menzies, Menzies' Journal of Vancouver's Voyage April to October 1792, ed. C. F. Newcomb, MD (Victoria: Legislative Assembly, 1923), ix and x.

133 Archibald's letter to mother and others: Archibald Menzies, Menzies' Journal of Vancouver's Voyage.

134 Archibald Menzies, 1790–1794. Journal of Archibald Menzies, Surgeon and Botanist on Board Discovery, 8. Available through: Adam Matthew, Marlborough, Empire Online. http://www.empire. amdigital.co.uk/Documents/Details/Journal of Archibald Menzies surgeon and botanist on board Discovery_, accessed 14 October 2021.

135 Letter from Lord Morton: https://nla.gov.au/nla. obj-223065583/view, retrieved 16 September 2021. Unfortunately, while Cook was a master navigator, he was neither a philosopher nor a miracle worker, and was not able to fully embrace Lord Morton's hints. In taking possession of the east coast of Australia in the name of the English Crown, Cook contradicted the advice offered by Morton.

136 Letter from Lord Grenville to the Admiralty dated 11 February 1791, Government of New South Wales Library, https://transcripts.sl.nsw.gov.au/page/ copy-letter-written-lord-grenville-lords-admiralty-29-march-1791-series-4907-no-0001, retrieved 18 August 2021.

137 James McCarthy, Monkey Puzzle Man, 74.

138 Marines = naval infantry (soldiers).

139 Pitt was sent back from Hawaii on 7 February 1794, aboard the supply ship Daedalus.

140 Journal entry, 17 March 1793. Archibald Menzies, Hawaii Nei 128 Years Ago, 114.

141 Journal entry, 1 March 1793. Archibald Menzies, Hawaii Nei 128 Years Ago, 88.

142 Journal entry, 1 March 1793. Archibald Menzies, Hawaii Nei 128 Years Ago, 88–89.

143 Journal entry, 17 March 1793. Archibald Menzies, Hawaii Nei 128 Years Ago, 114.

144 The incident occurred in May 1792 while Discovery was far away in the Pacific Northwest; Vancouver was not able to address the issue with King Kahekilli until the Discovery returned to Hawaii at the end of that summer.

145 Journal entries for 19–21 March, 1793. Archibald Menzies, Hawaii Nei 128 Years Ago, 119–124.

146 Journal entries for 19–21 March, 1793. Archibald Menzies, Hawaii Nei 128 Years Ago, 119–124.

147 Journal entry, 3 March 1793. Archibald Menzies, Hawaii Nei 128 Years Ago, 89–90.

148 Journal entry, 20 February 1793. Archibald Menzies, Hawaii Nei 128 Years Ago, 66.

149 Journal entry, 8 March 1793. Archibald Menzies, Hawaii Nei 128 Years Ago, 99.

150 Journal entry, 8 March 1793. Archibald Menzies, Hawaii Nei 128 Years Ago, 100.

151 Journal entry, 15 February 1793. Archibald Menzies, Hawaii Nei 128 Years Ago, 54–55.

152 Journal entry, 20 February 1793. Archibald Menzies, Hawaii Nei 128 Years Ago, 62.

153 Midshipman Thomas Manby estimated 30,000 people on hand for the ships' arrival.

154 Journal entry, 22 February 1793. Archibald Menzies, Hawaii Nei 128 Years Ago, 68. In return, Vancouver presented Kamehameha with a pennant of the kind reserved exclusively for British war ships, to fly from the royal double canoe. He also directed the Discovery's carpenters to help build a 9 metre long boat for Kamehameha.

155 Journal entry, 5 March 1793. Menzies, Archibald Menzies, Hawaii Nei 128 Years Ago, 94.

156 James Lee (1715–1795) was born in Selkirk, worked at the Chelsea Physic Garden, was a correspondent of Carl Linnaeus, and a partner with Lewis Kennedy in the Vineyard Nursery of Hammersmith.

157 Journal entry, 4 March 1792. Archibald Menzies, Hawaii Nei 128 Years Ago, 19.

158 Journal entry, 25 Feb 1793. Archibald Menzies, Hawaii Nei 128 Years Ago, 80–81.

159 Journal entry from 25 February 1793. Archibald Menzies, Hawaii Nei 128 Years Ago, 79.

160 Journal entry, 25 February 1793. Archibald Menzies, Hawaii Nei 128 Years Ago, 79.

161 Journal entry, 25 February 1793. Archibald Menzies, Hawaii Nei 128 Years Ago, 83.

162 Journal entry, 25 February 1793. Archibald Menzies, Hawaii Nei 128 Years Ago, 83.

163 Journal entry, 14 January 1794. Archibald Menzies, Hawaii Nei 128 Years Ago, 146.

164 Journal entry, 18 January 1794. Archibald Menzies, Hawaii Nei 128 Years Ago, 157.

165 Journal entry, 18 January 1794. Archibald Menzies, Hawaii Nei 128 Years Ago, 157.

166 Journal entry, 16 January 1794. Archibald Menzies, Hawaii Nei 128 Years Ago, 149.

167 Journal entry, 2 February 1794. Archibald Menzies, Hawaii Nei 128 Years Ago, 174.

168 Journal entry, 25 January 1794. Archibald Menzies, Hawaii Nei 128 Years Ago, 171–172.

169 Journal entry, 25 January 1794. Archibald Menzies, Hawaii Nei 128 Years Ago, 171–172.

170 The thermometer was given to Archibald by the Commander of the Cape Colony, Col. R. J. Gordon. An accomplished adventurer and explorer of Scottish descent, Gordon was also multilingual and fluent in English, French, Dutch, and the Indigenous African languages KhoeKhoe and Xhosa.

171 Journal entry, 12 February 1794. Archibald Menzies, Hawaii Nei 128 Years Ago, 187.

172 Journal entry, 12February 1794. Archibald Menzies, Hawaii Nei 128 Years Ago, 188.

173 Journal entry, 14 February 1794. Archibald Menzies, Hawaii Nei 128 Years Ago, 190.

174 Journal entry, 15 February 1794. Archibald Menzies, Hawaii Nei 128 Years Ago, 196.

175 Journal entry, 15 February 1794. Archibald Menzies, Hawaii Nei 128 Years Ago, 196.

176 Journal entry, 15 February 1794. Archibald Menzies, Hawaii Nei 128 Years Ago, 196.

177 His barometer reading was remarkably accurate. It was later found that he was short by just 43 feet (13 metres). Although, in fairness, his journal notes did advise 'But it is necessary to observe that the correction for temperature of the atmosphere has not been allowed in this calculation … which will make some difference in the result of the observations'. It would be another 40 years before anyone climbed the peak again (David Douglas in 1834) to check his work. Archibald was certainly the first European to reach the peak, and some think he was the first person ever to do so.

178 Journal entry, 16 February 1794. Archibald Menzies, Hawaii Nei 128 Years Ago, 198.

179 Journal entry, 16 February 1794. Archibald Menzies, Hawaii Nei 128 Years Ago, 199.

180 Journal entry, 16 February 1794. Archibald Menzies, Hawaii Nei 128 Years Ago, 199.

181 Journal entry, 21 April 1792. Archibald Menzies, Menzies' Journal of Vancouver's Voyage.

182 Journal entry, 27 April 1792. George Vancouver, A Voyage of Discovery to the North Pacific Ocean and Round the World, Vol 1 (London: G. G. and J. Robinson; J. Edwards, 1798), 210.

183 George Vancouver, A Voyage of Discovery to the North Pacific , Vol 1, 33.

184 George Vancouver, A Voyage of Discovery to the North Pacific , Vol 1, 32.

185 The small island in the Columbia River, located between the present-day cities of Portland, Oregon and Vancouver, Washington was named 'Menzies Island' by Lt Broughton. American explorers Lewis and Clark re-discovered the island ten years later and re-named it Canoe Island. It is now known as Hayden Island.

186 George Vancouver, A Voyage of Discovery to the North Pacific, Vol 1, 34.

187 Journal entry, 29 April 1792. Archibald Menzies, Menzies' Journal of Vancouver's Voyage, 48. Captain Gray, after meeting with Archibald and Puget in April of 1792, returned to the Columbia River area and crossed the intimidating bar at its mouth. But, since he was only interested in the fur trade and did not see much prospect for that here, he did not venture further upstream. He named the river after his ship, Columbia.

188 Journal entry, 26 May 1792. Archibald Menzies, Menzies' Journal of Vancouver's Voyage, 77.

189 Journal entry, 28 May 28 1792. Archibald Menzies, Menzies' Journal of Vancouver's Voyage, 78.

190 Journal entry, 1 May 1792. Archibald Menzies, Menzies' Journal of Vancouver's Voyage, 51.

191 If he had been able to reach the summit of Mount Baker, Archibald would have measured its height at 10,781 feet [3,286 metres] … about 3,000 ft [914 metres] lower than Mauna Loa.

192 Journal entry, 12 June 1792. Archibald Menzies, Menzies' Journal of Vancouver's Voyage, 54.

193 In his journal Vancouver said the area '… afforded not a single prospect that was pleasing to the eye, the smallest recreation on shore, nor animal nor vegetable food, excepting a very scanty proportion of those eatables already described, and of which the adjacent country was soon exhausted, after our arrival'.

194 Journal entry, 2 July 2, 1792. Archibald Menzies, Menzies' Journal of Vancouver's Voyage, 112). The hill is known today as Nipple Summit (2,876 feet) and is located just north of Cortes Island.

195 Journal entry, 2 July 1792. Archibald Menzies, Menzies' Journal of Vancouver's Voyage, 112.

196 Presumably this is Cassel Falls on West Redonda Island.

197 Journal entry, 2 July 1792. Archibald Menzies, Menzies' Journal of Vancouver's Voyage, 99.

198 Journal entry, 13 July 1792. Archibald Menzies, Menzies' Journal of Vancouver's Voyage, 83.

199 Journal entry, 13 July 1792. Archibald Menzies, Menzies' Journal of Vancouver's Voyage, 82.

200 Journal entry, 13 July 1792. Archibald Menzies, Menzies' Journal of Vancouver's Voyage, 83.

201 Journal entry,13 July 1792. Archibald Menzies, Menzies' Journal of Vancouver's Voyage, 83.

202 Journal entry, 12 June 1792. Archibald Menzies, Menzies' Journal of Vancouver's Voyage, 54.

203 Journal entry, 12 July 1792. Archibald Menzies, Menzies' Journal of Vancouver's Voyage, 117.

204 Archibald would later note in his journal (2 October 1792) that 'saluting was so common among the Trading Vessels that visited the Cove that there was scarcely a day past without puffings of this kind from some Vessel or other, & we too followed the example, & puffed it away as well as any of them'.

205 Journal entry, 22 Sept 1792. Archibald Menzies, Menzies' Journal of Vancouver's Voyage, 124.

206 Tahsis was, as the crew of Discovery and Chatham had begun to suspect from their cruise past Cape Mudge and through Johnstone's Strait, the key to an overland trade route connecting Nootka to the east coast (and other parts) of the island. Maquinna's control and oversight of the trade made him extremely important politically, and economically.

207 Journal entry, 5Sept 1792. Archibald Menzies, Menzies' Journal of Vancouver's Voyage, 117–118.

208 Letter to Banks, 14 Jan 1793, from Monterrey Bay, https://legalanswers.sl.nsw.gov.au/banks/section-11/series-61/61-16-letter-received-by-banks-from-archibald, retrieved 19 November 2021.

Archibald thought the island ought more rightly to have been named after the King: 'It should, I think, with more propriety be named after his Majesty as the name of King George's Sound is now extinct, which Capt. Cook first named for him'.

209 Journal entry, 12 Oct 1792. Archibald Menzies, Menzies' Journal of Vancouver's Voyage, 130–131.

210 Clive L. Justice, Mr. Menzies' Garden Legacy: An Illustrated History (Vancouver: Big Leaf Maple Books, 2017), 81.

211 Journal entry, 6 June 1792. Archibald Menzies, Menzies' Journal of Vancouver's Voyage, 48.

212 Named after Rear-Admiral John Jervis (1735–1823) by George Vancouver in 1792.

213 Journal entry, 23 June 1792. Archibald Menzies, Menzies' Journal of Vancouver's Voyage, 62.

214 Correspondence to Sir Joseph Banks, 29 April 1794, https://www.sl.nsw.gov.au/banks/section-11/series-61/61-18-letter-received-by-banks-from-archibald, retrieved 9 November 2021.

215 Cranstoun returned aboard Daedalus on 8 September 1792.

216 Letter from Archibald to Banks, 14 January 1793. From: James McCarthy, Monkey Puzzle Man, 142.

217 This happened on August of 1792, near Deserters' Island, off the north coast of Vancouver Island.

218 Journal entry, 6 August 1792. Archibald Menzies, Menzies' Journal of Vancouver's Voyage, 95.

219 Journal entry, 4 August 1792. Archibald Menzies, Menzies' Journal of Vancouver's Voyage, 94.

220 Quadra died less than a year later, in Mexico City, from what is now believed to have been a brain tumour.

221 James McCarthy, Monkey Puzzle Man, 147.

222 More commonly known as the hemlock spruce tree.

223 Journal entry, 31 July 1792. Archibald Menzies, Menzies' Journal of Vancouver's Voyage, 92.

224 George Vancouver, A Voyage of Discovery to the North Pacific, Vol. IV, 47.

225 Journal entry, June 27, 1792. Archibald Menzies, Menzies' Journal of Vancouver's Voyage, 67.

226 In his journal (21 Jan 1793) Vancouver wrote, 'Joseph Murgatroyd … was last observed opening the gun room ports, and whilst so employed, had probably been induced to seek his own destruction by contriving to let himself down into the sea since it was scarcely possible he could have met that fate there by accident'.

227 Journal entry, 8 September 1792. Archibald Menzies, Menzies' Journal of Vancouver's Voyage, 122.

228 George Vancouver, A Voyage of Discovery to the North Pacific, Vol. IV, 179.

229 George Vancouver, A Voyage of Discovery to the North Pacific, Vol. IV, 172.

230 George Vancouver, A Voyage of Discovery to the North Pacific, Vol. IV, 174. Whether any of the natives were killed is unknown. Vancouver wrote, 'To what degree our firing did execution, was not ascertained. Some of the natives were seen to fall, as if killed or severely wounded; and great lamentations were heard after they had gained their retreat in the woods, from whence they showed no disposition to renew their attack'.

231 George Vancouver, A Voyage of Discovery to the North Pacific, Vol. IV, 175–176.

232 George Vancouver, A Voyage of Discovery to the North Pacific, Vol. IV, 180.

233 Some of the other noteworthy persons he is known to have met include Mr Robert Duffin, formerly chief mate aboard Argonaut when it was under command of Meares and seized at Nootka by Martinéz. Duffin was now back as commander of a brig, sailing from Macao under cover of Portuguese colours. He also met American fur trader Mr McGee of the Margaret from Boston, and British trader Mr Brown ('the only English vessels who had an exclusive grant from government for trading on this coast') who was in charge of the Butterworth, Jackal, and Prince Lee Boo.

Had Archibald visited Bella Coola a few weeks later, he would have met overland explorer Alexander Mackenzie. Mackenzie was exploring major rivers, looking for the Northwest Passage or for routes that could expedite trade from west coast to east. Apart from their unquenchable thirst for outdoor adventure, the two would have had much in common to talk about: Mackenzie was born on the Isle of Lewis in the Hebrides, and Archibald had met many people from the Mackenzie clan while botanizing there 15 years earlier. Alexander's father, a veteran of the Jacobite rebellion, moved the family from Scotland to New York just before the American Revolutionary War but, like many Scots whom Archibald met there ten years later, they fled to the safety of British North America after the war ended. When adventurous, wandering, Alexander Mackenzie eventually arrived at Bella Coola he learned from the local Indigenous people, the Nuxalkmc, that two other Europeans – 'Macubah' (Vancouver) and 'Bensins' (Menzies) had recently visited them by boat.

234 Vancouver disembarked Discovery rather unceremoniously a few weeks earlier (13 September 1795) at Shannon, Ireland, leaving the last leg of the epic voyage in the care of Baker.

235 Archibald sent a copy of the letter to Banks, which can be read here: https://transcripts.sl.nsw.gov.au/page/letter-received-banks-archibald-menzies-1794-series6121-no-0001.

236 Letter from Archibald Menzies to Banks, 14 September 1795. From: James McCarthy, Monkey Puzzle Man, 167.

237 Letter from Archibald Menzies to Banks, 14 September 1795. From: James McCarthy. Monkey Puzzle Man, 166.

238 After learning from the British governor in St Helena that the war had been expanded to include Holland, the crew of Discovery captured a fully laden and slow-moving Dutch trade vessel (the Macassar). The Discovery returned to London not just with charts and plants but also a prize ship. In time of war, disrupting the enemy's economy had just as much impact as any purely military engagements, and was considered a vital part of naval warfare. Any merchant vessel sailing under enemy colours was forfeit so long as one could catch it, and the prize money was divided among the crew.

239 Coincidentally, Louis XVI was beheaded on the same day that one of Discovery's carpenters committed suicide by throwing himself overboard (21 January 1793).

240 The nitrous acid fumigation worked, not because it purified the air, but because it was effective against lice and killed bacteria. This detail would be discovered later.

241 George Spencer, the 2nd Earl Spencer, is the great-great-great grandfather of the 9th Earl Spencer, Charles Spencer, and his sister Diana, Princess of Wales.

242 James Carmichael Smyth, An Account of the Experiment Made at the Desire of the Lords Commissioners of the Admiralty, on board the Union Hospital Ship . . .. (London, 1796), v-vi.

243 A 'pipkin' is a type of earthenware cooking pot.

244 James Carmichael Smyth, An Account of the Experiment, 11–14.

245 James Carmichael Smyth, An Account of the Experiment, 16.

246 'Soil' is another word for faeces.

247 James Carmichael Smyth, An Account of the Experiment, 16–17.

248 James Carmichael Smyth, An Account of the Experiment, 21.

249 James Carmichael Smyth, An Account of the Experiment, 37.

250 James Carmichael Smyth, An Account of the Experiment, 43.

251 Vasily Chicagov had sailed the coast of Siberia in 1764 with scientist Mikhail Lomonosov, searching for a northern passage from the Atlantic to the Pacific. Educated in England and perfectly bilingual, Chicagov and Archibald undoubtedly discussed their experiences in the Bering Sea.

252 James Carmichael Smyth, An Account of the Experiment, 75.

253 Gilbert Blane and James Johnston, Letters Etc on the Subject of Quarantine, (St. George's Fields: Printed at the Philanthropic Reform, 1799), 24.

254 The recipient of Jenner's experiment was an eight-year-old boy, successfully inoculated with matter gathered from a dairymaid's cowpox lesions, on 14 May 1796.

255 This comment was communicated in a letter to Joseph Banks, 22 April 1797. Derrick Baxby, 'Edward Jenner's Unpublished Cowpox Inquiry and the Royal Society: Everard Home's Report to Sir Joseph Banks', Medical History 43, no. 1 (1999), 108–10, https://doi.org/10.1017/S0025727300064747.

256 The cartoon was printed by Mrs Humphrey's print-shop, located across the street from Vancouver's Bond Street home and was displayed in the shop window for all to see.

257 Archibald was appointed to HMS Princess Augusta effective 30 November 1796 and remained on her books until May of 1799. Some claims have been made that Archibald served on a Royal Yacht around this time, but likely some have confused the Princess Augusta yacht that Archibald served on with the Danish yacht HDMS Kronprindsens Lystfregat, launched in 1785 for the Prince Royal of Denmark, which was subsequently acquired by the British Navy and renamed Princess Augusta in 1818.

258 Napoleon took many scholars and artists with him when he invaded Egypt in 1798, and their

writings, collections, and illustrations captured the imaginations of many people not only in Paris but also in London and around the world. Napoleon had much success on land but less luck at sea. The Battle of the Nile, also in 1798, was a decisive victory for the British Royal Navy under Horatio Nelson and yet another event that made exotic Egypt a topical subject among the citizens of London.

259 This bizarre event is alleged to have occurred in the St James area of London on the evening of 17 February 1797.

260 Archibald's 3 January 1798 letter to Banks can be read in full here: https://www.sl.nsw.gov.au/banks/section-11/series-61/61-35-letter-received-by-banks-from-archibald.

261 The average life expectancy at the time was age forty, so Vancouver's passing at that same age was not particularly alarming. Although Archibald had attended to Vancouver as his personal physician aboard the Discovery, he had likely considered Vancouver's anxiety, irritability, and fatigue to be the outcome of his natural temperament combined with the pressures of command. Archibald would never have considered that Vancouver might be suffering from Graves' disease or possibly Addison's disease – since Graves' disease would not be identified as such for at least another twenty-five years, and Addison's would not be known as such until 1855 – though many today suspect that was the case.

262 The University of Aberdeen was formed in 1860 with the union of the two existing Colleges in Aberdeen, King's College and Marischal College. At this time, the MD degree was often awarded at both King's and Marischal Colleges by attestation – on the recommendations of two or three established physicians – rather than after completion of studies at Aberdeen.

263 The transition from HMS Princess Augusta to Sans Pareil was not seamless. Between the beginning of May and the end of July, 1799 Archibald was shuffled from the roster of Princess Augusta to Sans Pareil, to Tamar, and back to Sans Pareil again.

264 Birth and burial records indicate Adam was born 3 June 1776 and died 12 April 1828, aged 52. Ann was born 25 Dec 1779 and buried 8 Oct 1828, aged 49. Both were born at Vine Street in the Parish of Westminster. Ann's signature is on the marriage record of Archibald and Janet, as is the signature of John Walker.

265 A 1795 London directory includes an 'Adam Brown, coal merchant' living at 9 Ranelagh Street, Pimlico. This is not proven to be Janet's father but the author thinks it probable: the Pimlico address is less than 2 km from the Brompton Square address her brother Adam was known to be living in 1828, and is in the same parish where Janet and Archibald were married.

266 The author acknowledges a debt of gratitude to art historian and miniature portrait specialist Emma Rutherford for examining his photographs of the portraits held at the Royal BC Archives and identifying the artist and time period of creation. Until this discovery, the artist of the miniatures was unattributed.

267 Alexander Gordon, Dictionary of National Biography, 1855–1900, Volume 53. From: https://en.wikisource.org/wiki/Dictionary_of_National_Biography,_1885-1900/Smith,_Pleasance, retrieved 31 December 2021.

268 Letter from Archibald to James Edward Smith, 30 March 1805. Retrieved from: http://linnean-online.org/view/correspondence_by_collection/smith=5Fcorrespondence/Menzies=3AArchibald=3A=3A/Smith=3A_Sir_James_Edward=3A=3A.html.

269 Letter from William Fitt Drake to James Edward Smith, 19 April 1805. Retrieved from: http://linnean-online.org/view/correspondence_by_collection/smith=5Fcorrespondence/Drake=3AWilliam_Fitt=3A=3A/Smith=3A_Sir_James_Edward=3A=3A.html.

270 Letter from William Fitt Drake to James Edward Smith, 19 April 1805. Retrieved from: http://linnean-online.org/view/correspondence_by_collection/smith=5Fcorrespondence/Drake=3AWilliam_Fitt=3A=3A/Smith=3A_Sir_James_Edward=3A=3A.html.

271 Pierce Egan, Pierce Egan's Book of Sports, and Mirror of Life, (1832), 228.

272 Pierce Egan, Pierce Egan's Book of Sports, 228.

273 King James Bible. Proverbs 16:18, retrieved from: https://www.kingjamesbibleonline.org/Proverbs-16-18/ .

274 Retrieved 27 December 2021 from: https://threedecks.org/index.php?display_type=show_battle&id=157.

275 Oregon History Project, https://www.oregonhistoryproject.org/articles/historical-records/under-attack-at-nootka-sound-1803/#.Yci24xPMLeo, retrieved 26 December 2021.

276 John Rodgers Jewitt and Richard Alsop, The Adventures and Sufferings of John R. Jewitt, Only Survivor of the Ship Boston, During a Captivity of Nearly Three Years among the Savages of Nootka Sound: With an Account of the Manners, Mode of Living and Religious Opinions of the Natives. From: https://www.canadiana.ca/view/oocihm.27869, retrieved 25 December 2021.

277 This was confirmed through personal correspondence between the author and the Royal Archives at Windsor Castle, 6 July 2021.

278 J. J. Keevil, Archibald Menzies, 1754–1842, 807–808.

279 'Rhio-Rhio' is a version of the more correct name, Liholiho.

280 'Tamehemeo' is a version of the more correct name: Kamehameha.

281 David W. Forbes, ed., Hawaiian National Bibliography, 1780–1900: Vol 1: 1780–1830. (Honolulu: University of Hawaii Press, 1999), 407. https://books.google.ca/books?id=WA3pXblqS20C&printsec=frontcover&source=gbs_ge_summary_r&cad=0#v=onepage&q&f=false, retrieved 28 December 2021.

282 A vaccine for measles would not be available until the 1960s.

283 Napoleon's dramatic reign as emperor came to an end three years earlier, and Louis XVIII – who had been living in exile in England since 1807, with the support of George III – had returned to the French throne. It was safe for Englishmen like Johnstone to live in Paris again.

284 Johnstone's will, which lists Archibald and Adam Brown (from Chatham) as executors, can be found here: https://discovery.nationalarchives.gov.uk/details/r/D152404.

285 Adam left a sum of £ 500 to Janet in his will.

286 David Douglas, 'Sketch of a Journey to the Northwestern Parts of the Continent of North America during the Years 1824–25–26–27' Quarterly of the Oregon Historical Society, 5, no. 3 (1904), 244, http://www.jstor.org/stable/20609621.

287 They would both be in good company. The Kensall Green Cemetery was just nine years old when Archibald was lowered into his grave, but it was already a prestigious neighbourhood. His and Janet's neighbours in death include children of George III: Prince Augustus Frederick, the Duke of Sussex, in 1843 and Princess Sophia in 1848; Scottish botanist (and Joseph Bank's last librarian), Robert Brown in 1858. Even today it is considered one of the most prestigious final resting places in London.

288 The SS Beaver was a hybrid model; she was powered by sail from London to the Columbia River and had her steam boilers and engines and paddles assembled there. She went on to become an iconic vessel in the history of British Columbia.

289 https://sourcebooks.fordham.edu/mod/carlyle-times.asp, retrieved 28 December 2021.

290 Highland dress was outlawed by George II in 1746, following the Jacobite Rebellion; the Dress Act was repealed by George III in 1782. Queen Victoria and Prince Albert have been credited with making Highland dress stylish.

291 A Chinook Wawa dictionary was published in 1863. Approximately 100,000 people could speak Chinook Wawa in 1875, and it was used widely in court testimony, newspaper advertising, and missionary activity among Indigenous peoples and in everyday conversation from central British Columbia to northern California.

292 Information obtained from exhibits at the UBC Museum of Anthropology, June 2021.

293 The ban would not be lifted until 1951. Most of the items confiscated at the incident mentioned have been returned and are now housed in the U'mista Cultural Centre at Alert Bay and the Nuyumbalees Cultural Centre at Cape Mudge. More information about the potlach ban can be found here: https://umistapotlatch.ca/nos_masques_come_home-our_masks_come_home-eng.php .

294 Hon. Jody Wilson-Raybould, PC, KC, was appointed Minister of Justice and Attorney General of Canada in 2015. She is a member of the We Wai Kai Nation.

295 Bruce Cumings, Dominion from Sea to Sea: Pacific Ascendancy and American Power, (Yale University Press, 2009), 180. According to the 2010 US Census, 371,000 people in Hawaii self-identified as 'Native Hawaiian'.

296 Lewis and Clark, during their expedition of 1804–1806, explored the Columbia River from its confluence with the Snake River near Pasco, Washington, downstream to its mouth at present day Astoria. But from 1806 to 1811, it was British surveyor David Thompson who unravelled the mystery of the remaining unknown three-quarters of the Columbia's course.

297 Isidor Lowenstern, 'Ascent of Mauna Loa, 6th February 1839', transcribed from Royal Geographical Society Archives Journal MS 1841 Lowenstern, https://people.csail.mit.edu/bkph/articles/Isidor_Loewenstern.pdf, retrieved January 12, 2022.

298 Hooker was director of Kew Gardens from 1831–1865 and also a friend of Joseph Banks.

299 Letter from David Douglas to W. J. Hooker, 6 May 1834. From: James McCarthy, Monkey Puzzle Man, 181.

300 These items were acquired by Lieutenant Colonel F. R. S. Balfour, who donated them to the Royal BC Museum in 1943 and where, as a result of the research conducted for this book, the watch was recently rediscovered. The existence and location of the display table was unknown until traced by the author, in July 2023, to the Scottish home of one of Balfour's descendants.

301 The duplicate bust was commissioned by Hon Ted Menzies, PC, of Alberta, in collaboration with the author, and with the support and contributions of many friends in Canada, Australia, New Zealand, USA, and elsewhere.

INDEX

People

Places

Ships